Francis Warner

Physical Expression

Its modes and principles

Francis Warner

Physical Expression
Its modes and principles

ISBN/EAN: 9783337312404

Printed in Europe, USA, Canada, Australia, Japan

Cover: Foto ©berggeist007 / pixelio.de

More available books at **www.hansebooks.com**

THE INTERNATIONAL SCIENTIFIC SERIES.
VOLUME LI.

THE

INTERNATIONAL SCIENTIFIC SERIES.

EACH BOOK COMPLETE IN ONE VOLUME, 12MO, AND BOUND IN CLOTH.

I. FORMS OF WATER: a Familiar Exposition of the Origin and Phenomena of Glaciers. By J. TYNDALL, LL. D., F. R. S. With 25 Illustrations. $1.50.

II. PHYSICS AND POLITICS; Or, Thoughts on the Application of the Principles of "Natural Selection" and "Inheritance" to Political Society. By WALTER BAGEHOT. $1.50.

III. FOODS. By EDWARD SMITH, M. D., LL. B., F. R. S. With numerous Illustrations. $1.75.

IV. MIND AND BODY: The Theories of their Relation. By ALEXANDER BAIN, LL. D. With Four Illustrations. $1.50.

V. THE STUDY OF SOCIOLOGY. By HERBERT SPENCER. $1.50.

VI. THE NEW CHEMISTRY. By Professor J. P. COOKE, of Harvard University. With 31 Illustrations. $2.00.

VII. ON THE CONSERVATION OF ENERGY. By BALFOUR STEWART, M. A., LL. D., F. R. S. With 14 Illustrations. $1.50.

VIII. ANIMAL LOCOMOTION; or, Walking, Swimming, and Flying. By J. B. PETTIGREW, M. D., F. R. S., etc. With 130 Illustrations. $1.75.

IX. RESPONSIBILITY IN MENTAL DISEASE. By HENRY MAUDSLEY, M. D. $1.50.

X. THE SCIENCE OF LAW. By Professor SHELDON AMOS. $1.75.

XI. ANIMAL MECHANISM: A Treatise on Terrestrial and Aërial Locomotion. By Professor E. J. MAREY. With 117 Illustrations. $1.75.

New York: D. APPLETON & CO., 1, 3, & 5 Bond Street.

The International Scientific Series.—(Continued.)

XII. THE HISTORY OF THE CONFLICT BETWEEN RELIGION AND SCIENCE. By J. W. DRAPER, M. D., LL. D. $1.75.

XIII. THE DOCTRINE OF DESCENT AND DARWINISM. By Professor OSCAR SCHMIDT (Strasburg University). With 26 Illustrations. $1.50.

XIV. THE CHEMICAL EFFECTS OF LIGHT AND PHOTOGRAPHY. By Dr. HERMANN VOGEL (Polytechnic Academy of Berlin). Translation thoroughly revised. With 100 Illustrations. $2.00.

XV. FUNGI: Their Nature, Influences, Uses, etc. By M. C. COOKE, M. A., LL. D. Edited by the Rev. M. J. BERKELEY, M. A., F. L. S. With 109 Illustrations. $1.50.

XVI. THE LIFE AND GROWTH OF LANGUAGE. By Professor WILLIAM DWIGHT WHITNEY, of Yale College. $1.50.

XVII. MONEY AND THE MECHANISM OF EXCHANGE. By W. STANLEY JEVONS, M. A., F. R. S. $1.75.

XVIII. THE NATURE OF LIGHT, with a General Account of Physical Optics. By Dr. EUGENE LOMMEL. With 188 Illustrations and a Table of Spectra in Chromo-lithography. $2.00.

XIX. ANIMAL PARASITES AND MESSMATES. By Monsieur VAN BENEDEN. With 83 Illustrations. $1.50.

XX. FERMENTATION. By Professor SCHÜTZENBERGER. With 28 Illustrations. $1.50.

XXI. THE FIVE SENSES OF MAN. By Professor BERNSTEIN. With 91 Illustrations. $1.75.

XXII. THE THEORY OF SOUND IN ITS RELATION TO MUSIC. By Professor PIETRO BLASERNA. With numerous Illustrations. $1.50.

New York: D. APPLETON & CO., 1, 3, & 5 Bond Street.

XXIII. STUDIES IN SPECTRUM ANALYSIS. By J. Norman Lockyer, F. R. S. With Six Photographic Illustrations of Spectra, and numerous Engravings on Wood. $2.50.

XXIV. A HISTORY OF THE GROWTH OF THE STEAM-ENGINE. By Professor R. H. Thurston. With 163 Illustrations. $2.50.

XXV. EDUCATION AS A SCIENCE. By Alexander Bain, LL. D. $1.75.

XXVI. STUDENTS' TEXT-BOOK OF COLOR, or, Modern Chromatics. With Applications to Art and Industry. By Professor Ogden N. Rood, Columbia College. New edition. With 130 Illustrations. $2.00.

XXVII. THE HUMAN SPECIES. By Professor A. de Quatrefages, Membre de l'Institut. $2.00.

XXVIII. THE CRAYFISH: an Introduction to the Study of Zoölogy. By T. H. Huxley, F. R. S. With 82 Illustrations. $1.75.

XXIX. THE ATOMIC THEORY. By Professor A. Wurtz. Translated by E. Cleminshaw, F. C. S. $1.50.

XXX. ANIMAL LIFE AS AFFECTED BY THE NATURAL CONDITIONS OF EXISTENCE. By Karl Semper. With Two Maps and 106 Woodcuts. $2.00.

XXXI. SIGHT: An Exposition of the Principles of Monocular and Binocular Vision. By Joseph Le Conte, LL. D. With 132 Illustrations. $1.50.

XXXII. GENERAL PHYSIOLOGY OF MUSCLES AND NERVES. By Professor J. Rosenthal. With 75 Illustrations. $1.50.

XXXIII. ILLUSIONS: A Psychological Study. By James Sully. $1.50.

XXXIV. THE SUN. By C. A. Young, Professor of Astronomy in the College of New Jersey. With numerous Illustrations. $2.00.

New York: D. APPLETON & CO., 1, 3, & 5 Bond Street.

XXXV. VOLCANOES: What they Are and what they Teach. By JOHN W. JUDD, F. R. S., Professor of Geology in the Royal School of Mines. With 96 Illustrations. $2.00.

XXXVI. SUICIDE: An Essay in Comparative Moral Statistics. By HENRY MORSELLI, M. D., Professor of Psychological Medicine, Royal University, Turin. $1.75.

XXXVII. THE FORMATION OF VEGETABLE MOULD, THROUGH THE ACTION OF WORMS. With Observations on their Habits. By CHARLES DARWIN, LL. D., F. R. S. With Illustrations. $1.50.

XXXVIII. THE CONCEPTS AND THEORIES OF MODERN PHYSICS. By J. B. STALLO. $1.75.

XXXIX. THE BRAIN AND ITS FUNCTIONS. By J. LUYS. $1.50.

XL. MYTH AND SCIENCE. By TITO VIGNOLI. $1.50.

XLI. DISEASES OF MEMORY: An Essay in the Positive Psychology. By TH. RIBOT, author of "Heredity." $1.50.

XLII. ANTS, BEES, AND WASPS. A Record of Observations of the Habits of the Social Hymenoptera. By Sir JOHN LUBBOCK, Bart., F. R. S., D. C. L., LL. D., etc. $2.00.

XLIII. SCIENCE OF POLITICS. By SHELDON AMOS. $1.75.

XLIV. ANIMAL INTELLIGENCE. By GEORGE J. ROMANES. $1.75.

XLV. MAN BEFORE METALS. By N. JOLY, Correspondent of the Institute. With 148 Illustrations. $1.75.

XLVI. THE ORGANS OF SPEECH AND THEIR APPLICATION IN THE FORMATION OF ARTICULATE SOUNDS. By G. H. VON MEYER, Professor in Ordinary of Anatomy at the University of Zürich. With 47 Woodcuts. $1.75.

XLVII. FALLACIES: A View of Logic from the Practical Side. By ALFRED SIDGWICK, B. A., Oxon. $1.75.

XLVIII. ORIGIN OF CULTIVATED PLANTS. By ALPHONSE DE CANDOLLE. $2.00.

XLIX. JELLY-FISH, STAR-FISH, AND SEA-URCHINS. Being a Research on Primitive Nervous Systems. By GEORGE J. ROMANES. $1.75.

L. THE COMMON SENSE OF THE EXACT SCIENCES. By the late WILLIAM KINGDON CLIFFORD. $1.50.

New York: D. APPLETON & CO., 1, 3, & 5 Bond Street.

THE INTERNATIONAL SCIENTIFIC SERIES

PHYSICAL EXPRESSION

ITS MODES AND PRINCIPLES

BY

FRANCIS WARNER, M.D. Lond., F.R.C.P.

ASSISTANT PHYSICIAN, AND LECTURER ON BOTANY TO THE LONDON
HOSPITAL; FORMERLY PHYSICIAN TO THE EAST
LONDON HOSPITAL FOR CHILDREN

WITH FIFTY-ONE ILLUSTRATIONS

NEW YORK
D. APPLETON AND COMPANY
1, 3, AND 5 BOND STREET
1886

PREFACE.

This work is addressed to those who are interested in studying Man as a living and thinking being.

Mind is the highest faculty of man; what Mind is we do not know, and probably we cannot know, but there is abundant evidence that mind is in some way connected with brain-action. In man, the manifestations of mind are very slight in those whose brains are small and ill formed; defective action of the mind is often found to be coincident with physical disease of the brain; injury to the brain may deprive a man of his mental power. In animals, the signs of intelligence are in some degree proportional to the size and structure of the brain. A strong line of argument on this point has been put forward by Dr. Bastian.*

I have found it convenient to use the term "mentation" for that physical action of the brain which is associated with the phenomena of mind; as thus defined, mentation is a function of the brain, physical in kind, and capable of physical investigation.

In the arguments here used it is postulated as a working hypothesis, that all physical phenomena are due to physical causes, or necessarily follow upon certain physical antecedents, and that every physical change is due to a purely physical force.

The methods here employed are those used in physical research; forces are studied in the bodies or material objects where they are seen, the physical expression of such forces being noted as an index of the actual but invisible.

---

* "Brain as an Organ of the Mind" (International Science Series), p. 188.

In looking at any example, or mode of expression, processes of analysis are used till we get at such elements of the phenomena as are capable of physical inquiry, and whose relations may be demonstrated.

Biology, medicine, and philosophy, with their working hypotheses, are drawn upon for facts, analogies, and arguments. I have aimed at practical results rather than intellectual satisfaction alone, hoping thus to assist those engaged in scientific work and others engaged in social life.

The general scheme and purport of this work may be gathered from the first chapter; the more practical and illustrative descriptions are contained in chaps. viii. to xiv. Summaries are appended to certain of the chapters (see index).

The chapter on Art Criticism illustrates the principles and arguments used throughout the work: as this may possibly be read by some without the preceding chapters, the principles of analysis of expression are here repeated;

they can be passed over by the reader who became acquainted with them in the earlier chapters.

The index has been rendered as complete as possible to enable such cross references to be made as may be required for a complete study of the scheme of the whole work.

F. W.

24, Harley Street, W.

CONTENTS.

CHAPTER I.

INTRODUCTORY 1

CHAPTER II.

EXPRESSION.

The term "expression" explained—Speech expressive of life; speech a criterion of life—The uniform coexistence of two phenomena makes the one an expression of the other—Abstract properties, and their objective signs; criteria of mind—Comparison of an idiot, and an intelligent man—Certain sounds and a tracing of the pulse express the action of the heart—Signs due to an afferent force; action of the sight of an object upon children; a thermometer indicates heat; the sensitive flame—The telephone exhibits impressionability; a receptive part, an expressive part, and an additional force thrown in—The phonograph exhibits localized impressionability, which is permanent—Impressionability and retentiveness in a child—Expression, direct and empirical—Impressionability and retentiveness—Nutrition; its signs—Movements in plants; by pulvini, by unequal growth—Expression by form, colour, temperature—Vital processes can only be studied by their expression; the importance of appreciating this in biological work—Summary 11

CHAPTER III.

EXPRESSION IN MAN AND IN ANIMALS.

The fact of expression does not prove vitality—In living things expression is the outcome of nutrition—Nutrition not considered here, only its expression—Expression when the outcome of processes in the subject is called direct—Growth results from local nutrition; such processes are termed trophic—Nutrition is an expression of life—Permanent impressionability; it is opposed to evolution, it may be expressed by reflex action—Retentiveness need not be permanent—Development and reflex action as modes of expression—Apparatus for reflex movement; reflexes may be congenital, or acquired—Reflected action, as expression by form or colour—Any outcome of function may be expressive—Expression by colour, sound, change of function in a part—Movements of an actor in anger—Trophic action illustrated by the growth of crystals, growth of the body—Coincident development of parts; coincident defects in imbeciles—Properties demonstrated by external forces—Heredity—Expression of the emotions—Summary 31

CHAPTER IV.

MODES OF EXPRESSION BY MOVEMENTS, AND THE RESULTS OF MOVEMENT.

Movement a physical and visible action; it is often observed in physiological inquiries; it is correlatable with other modes of force—A movement expresses the action that produces it—Examples of expression by movement: anger, laughter—Results of movement—Expression by the voice, apparatus of porcupines, stamping of rabbits—Secondary movements—Work done the result of movement—Posture as a result of movement—Subsidence of movement in sleep, in fatigue, and when the attention is attracted—Spontaneous and voluntary movements—Movements of a bee from flower to flower—Summary ... 48

CHAPTER V.

MOVEMENTS AND THE RESULTS OF MOVEMENTS CONSIDERED IN THE ABSTRACT, OR APART FROM WHAT THEY EXPRESS.

PAGE

Movements are means of expression—Movements classified as reflex, voluntary, spontaneous—The attributes of a movement are its quantity, kind, and time—Time of a movement most conveniently recorded by the graphic method—Frequency and duration, the importance of considerations as to time—Two movements, considered in relation to time, may be synchronous; this may depend upon an organic union of the motors, or upon each working in similar rhythm—Expression may consist of coincidences or combinations of movements—The number of possible combinations of synchronous movements of n subjects is $2n$, the number of sequences of such combinations is unlimited—Actions described as a series of movements—Description of a dog in terms of movement and growth—Co-ordinated and inco-ordinated movements—Walking described as a series of movements—Movements of an aggregation of independent individuals—Principles of analysis of movements—Description in anatomical terms—Contrast of movements of small parts and large parts of the body in their physiological significance—Inter-differentiation—Collateral differentiation of parts—Symmetry of movements, indicating like action on both sides of the brain—Asymmetry of movements common in the higher functions—Classification of movements: according to anatomical analysis; according to the physiological principles of analysis given above; as intelligent and non-intelligent; as synchronous or non-synchronous; as occurring in regular series; as accompanied by feelings, other classifications are suggested—Summary 66

CHAPTER VI.

PHYSIOLOGY OF EXPRESSION.

Modes of movement in plants; in the amœba; the ascidian has a nerve-mechanism, and apparatus for reflex movements—Nerve-mechanism of vertebrates—Nervo-muscular apparatus; nerve-muscular action—Do certain nerve-centres produce certain movements?—Ferrier's experiments — Cerebral localization — Nerve-centres — Visual perception indicated by movements—Time requisite for a reflex movement—Inhibition of movement—Physiological effects of light, in man, in plants—Light stimulates trophic and kinetic action—Effects of light in the new-born infant; movements stimulated, inhibited, co-ordinated—Retentiveness to effects of light—The brain of an idiot not thus impressionable to light—Summary of the effects of light—Extrinsic stimuli, mediate and immediate—Trophic action of light—Summary of effects of light on plants 82

CHAPTER VII.

PATHOLOGICAL FACTS AND EXPRESSION IN PATHOLOGICAL STATES.

Disease may destroy or irritate parts of the brain—Destruction of corpus striatum—Lateral deviation of the head and eyes—Effects of irritation in contrast with destruction of a brain area—Effects of disease on different sets of muscles—Facial palsy—Localization of disease—Epilepsy—Chorea—Analogy to movements in plants—Experiments with the mimosa—The study of chorea—Finger-twitching in nervous children—Tooth-grinding—Headaches in children; the physical signs—Cases of athetosis—Defects of development; their frequent coincidence ... 101

CHAPTER VIII.

POSTURES CONSIDERED AS MEANS OF EXPRESSION.

Definition of a posture—Simplicity of study—Historical records of postures—Postures of all parts—A change of posture is movement—A posture is due to resultant action of muscles and their nerve-centres—It is a direct mode of expression—Free or disengaged parts most expressive—A limb labouring is not susceptible to mental expression—Organic postures, as from difficult breathing—Postures due to gravity—Effect of gravity on plants—Gravity acts differently during sleep—It can affect the postures of the face—Postures due to reflex action—Spontaneous postures—Fallacies—Classification and analysis of postures—Coincident postures—Symmetry—Postures in art—Postures in animals—Postures in plants—Summary 140

CHAPTER IX.

POSTURES OF THE UPPER EXTREMITY.

Method of examination—Anatomy—The convulsive hand contrasted with the hand in fright—The feeble hand and the hand in rest—The straight extended hand, normal—Application of the principles of analysis—Straight extended hand with the thumb drooped—The nervous hand, seen in art—Energetic hand the antithesis of the nervous hand—Table giving analysis of postures, and application of the principles—Principles of analysis—Anatomical analysis—Small parts contrasted with large parts—Interdifferentiation — Collateral differentiation — Symmetry Excitation of weak centres—General excitement or weakness—Analogy—Antithesis—Fallacies—Methods of determining whether a posture is the outcome of the spontaneous action of the nerve-centres 154

CHAPTER X.

EXPRESSION IN THE HEAD.

Positions and movements of the head defined—Flexion the only symmetrical movement—Action of light in causing head movements; varying effect of such stimulus in different brain conditions—A weak posture—Effect of gravity—The head usually free—Application of the principles of analysis to head postures—Movements of the jaw—Physiognomy, or certain forms of the head—Summary 182

CHAPTER XI.

EXPRESSION IN THE HUMAN FACE.

The face as an index of the mind—Definition of the face; its structure—Facial muscles and their nerve supply; the sympathetic nerve—Form, colour, and mobile conditions of the face—Direct expression of brain action in the face; expression by coincident development—Action of the facial muscles—Method of analyzing a face: the upper, middle, and lower zones; symmetry; analysis of the expression of anxiety—Expression by trophic signs, skin—The intellectual and the vulgar face—The necessity of considering nerve-muscular signs as well as permanent conditions—Faces of idiots—Nutrition of the face; a dull and a bright face—What may be seen in a man's face—Impressions of previous movements of the face—Mental suffering compared with bodily suffering—Asymmetrical expressions; winking, snarling—The long face, due to paralysis of the nerve on one or both sides: due to mental states; facial palsy from brain disease—Intellectuality of facial movements—The face in fatigue; the expression of headache—The disengaged face free for mental expression—Cases of expression from brain disease —Conflict of muscles in the face 193

CHAPTER XII.

EXPRESSION IN THE EYES.

The eyeballs: their position and the mechanism for their movement—Iris, a muscular apparatus; its nerve supply —The pupil contracted by light, and accommodation for near vision; its reflex dilatation, and its variation in conditions of emotion and on brain stimulation—Mechanism of the eyelids—Importance of distinguishing expression by the eye and the parts around—Movements of the eyes — Loss of associated movements of eyes under chloroform, and in deep sleep—Movements of eyes from brain stimulation—Horizontal movements most common—Attraction and repulsion of the eyes by sight of an object —Spontaneous movements—Eyes free or disengaged— Mental states expressed by attraction or repulsion of the eyes—Horizontal and vertical movements contrasted— Intellectuality of upward movements 214

CHAPTER XIII.

EXPRESSION OF GENERAL CONDITIONS OF THE BRAIN AND OF THE EMOTIONS.

Expression of consciousness—Sleep—Fatigue—Exhaustion—
—Irritability—Nutrition—Rest—Activity—Impressionability—Expression of instinct and mentation—Expression of pain—The emotion of the beautiful 225

CHAPTER XIV.

EXPRESSION OF MIND IN THE INFANT AND ADULT.

Materialistic questions only entertained—The criteria of mind, what are they?—Physical study of signs of mind from infancy upwards—A subjective condition is only known to us by its physical expression—Brain properties necessary to mentation—Impressionability—Retentiveness—Relation of outcomings to afferent stimulus—Comparison of an infant with the adult, and an idiot with a healthy child—Description of an infant: its development and signs of potentiality—Impressionability: its attributes; delayed expression of impressions—Modes of expression are criteria of mind—Expression of distress—Memory—Subjective conditions studied by their expression—Thought 240

CHAPTER XV.

ANALYSIS OF EXPRESSION.

Analysis of the expression of fatigue—Localize the expression—Observe trophic signs, postures, movements—Analyze and classify movements according to the principles given—Analysis of Darwin's description of laughter, and Sir Charles Bell's description of joy—The importance of such analysis—Pope's description of Achilles—Study of a nervous subject—"A school inspection"—National modes of expression... 255

CHAPTER XVI.

CONSIDERATIONS AS TO THE ATTRIBUTES OF A PROPERTY OR FUNCTION—TIME, QUANTITY, KIND; AND AS TO THEIR RELATION.

The attributes of a property or function: time, quantity, kind —Attributes of the functions trophic and kinetic action, in one subject, in two or more subjects—Combinations and sequences of action—Proportional growth; equal proportional growth; similar development—Analogy between series of kinetic and trophic actions—Special combinations of action may result from afferent forces; this an important element in evolution—Heredity— Summary 267

CHAPTER XVII.

ART CRITICISM.

Art teaches the physiologist; all men can study expression —Bulwer's opinion—All expression of feeling is by nerve-muscular action—Importance of such studies—Expression of mental states—Hand and face specially indicative of the mind—Venus de' Medici, the nervous hand—Diana, the energetic hand—Composition—Etruscan drawing— Cain at Pitti Gallery, the hand in fright—The Dying Gladiator—Writings of Camper; his descriptions of expression — Antony Raphael Mengs — Study of nerve-muscular action—Weakness should not be expressed in place of beauty—The free hand—The object of figure composition in art—Fixed and mobile expression—Principles of analysis 289

CHAPTER XVIII.

LITERATURE.

Bulwer—Hartley—Gregory—Camper—Blanc—Marshall Hall —Tyndall—C. Darwin—Bibliography, with dates ... 321

CHAPTER XIX.

METHODS AND APPARATUS FOR OBTAINING GRAPHIC RECORDS OF MOVEMENTS IN THE LIMBS, ETC., AND ENUMERATING SUCH MOVEMENTS AND THEIR COMBINATIONS; PROBLEMS TO BE INVESTIGATED BY THESE METHODS.

Movement as a result of vital action is capable of physical experimentation—Early attempts to record movements—Apparatus described: the motor gauntlet; the recording tambours; the contact-making tambour; electrical counter; method of using the apparatus—Problems; as to muscular twitching in exhaustion—Movements of an infant—Inhibition by light—Measurement of differentiation of movements—Retentiveness—Signs of emotion—Potentiality for mind—Co-ordination—Athetosis—Chorea 347

LIST OF ILLUSTRATIONS.

FIG.		PAGE
1.	Longitudinal section of a pulvinus	25
2.	Chorea flexor movements	68
3.	Nervous system of an Ascidian	84
4.	Upper surface of the hemispheres of the monkey	86
5.	The left hemisphere of the monkey	87
6.	Upper surface of the human brain	88
7.	Lateral view of the human brain	89
8.	Hydrophobia. Head repelled by sight of water	98
9.	Tracing showing how the spontaneous movements of an infant were inhibited by sound and, again, by a strong light	101
10.	Right hemiplegia, with cerebral facial palsy, right side	108
11.	Left hemiplegia, with cerebral palsy, left side	109
12 and 13.	Tracing of involuntary movements of the finger in a nervous child	112
14.	Finger-tracings in chorea	113
15.	Tracings of movements in athetosis	128
16.	Cases of athetosis showing hand postures	130
17.	Dying Gladiator	145
18.	The convulsive hand	156
19.	The hand in fright	157
20.	The feeble hand	158
21.	The hand in rest	159
22.	The straight hand	160
23.	The straight extended hand with thumb drooped	161
24.	Hand intermediate between the hand in rest and the straight hand	161
25 and 33.	The nervous hand	163, 297

LIST OF ILLUSTRATIONS.

FIG.		PAGE
26.	The energetic hand	164
27.	Complete paralysis of the right side of the face	202
28 and 29.	Face of imbecile	205
30.	Paralysis agitans, in advanced stage	213
31.	Tracings of the spontaneous movements of an infant's hand during fifteen minutes	245
32.	Venus de' Medici	296
34.	Diana	298
35.	Feast of the gods	300
36.	Cain	302
37.	Hercules at rest	305
38.	A countenance perfectly placid	307
39.	Expressing surprise	307
40.	Contempt	308
41.	Complacency, friendliness, tacit joy	308
42.	Laughter	309
43.	Sorrow	309
44.	Weeping	310
45.	Diagram showing facial zones	320
46.	Motor gauntlet	349
47.	Motor gauntlet on hand	350
48.	Junctions for motor tubes	351
49.	Frame supporting the recording tambours, and electrical signals	352
50.	Contact-making tambours arranged in circuit	353
51.	Electrical counter	354

PHYSICAL EXPRESSION,

ITS MODES AND PRINCIPLES.

CHAPTER I.

INTRODUCTORY.

THIS work has been written as the outcome of observations made on children and adults, and it is hoped that, inasmuch as it is the result of observations on humanity, it may be found of some social use. The principles here put forward and illustrated have been applied very frequently to the consideration of questions of importance to individual men, women, and children, and this gives me some confidence that others may find practical results from these studies of the modes of expression. Children are the subjects most often referred to: and this is the result of many years' study of childhood. When I commenced the special study of children, it soon struck me that a sound and well-developed nerve-system was of primary importance as, firstly, giving vitality, and the power to endure organic diseases; and, secondly, on account of the great impressionability of the nerve-mechanism in childhood, and the immense importance of such impressions on the future moral and intellectual condition of the child.

In every scientific inquiry, whatever be the ultimate object in view, it is necessary to use the utmost accuracy as to method of procedure. The present work contains an analysis of some of the usual means of investigating the nerve-system, with a view to the establishment of an experimental method of inquiry as to the forces leading to its growth and development (see chap. xix.). This is a somewhat bold and ambitious project, and no one can be more conscious than the author of the difficulties that lie in the way. One special object in view is to show how much accuracy and clearness of insight may be obtained in biological work by dealing as much as possible with objective signs and physical forces only, always adhering to the postulate that every objective phenomenon must be the outcome of physical action.

My early studies were on the size, shape, and proportions of the head, as indicative of the brain within. Such observations led to but poor results; still, they were useful. The results of these observations are embodied in the account given of coincident development (see chap. xvi.).* I next set about considering how our knowledge of brain functions had been obtained, and analyzed the methods employed in such physiological work; and

---

* Mr. Charles Roberts's excellent manual on anthropometry gives much valuable information on this subject.

soon I saw that the principal methods of making physiological, and clinical, observations of brain conditions were derived from noting its motor-functions. Then, noting in all cases the motor outcome of brain action, I studied the spontaneous postures, and analyzed them.

In making such observations the spontaneous movements of the subjects attracted attention; they were difficult to analyze, but still it was obvious that, as physical phenomena, they were capable of analysis, record, and classification. An experimental method was then devised by which movements could be recorded with the aid of the graphic method described in chap. xix. The tracings thus obtained* were submitted to the criticism of mathematical and statistical friends, to whose help I am greatly indebted. Their remarks made it obvious that time, frequency, combinations, and sequences in series, were the attributes that should be noted in studying movements. Such knowledge as to method could not have been obtained without experiment. Following on these considerations, I founded the most important of the principles for the analysis of movements. The other principles which it is sought to establish are the result of analyzing recognized clinical modes of procedure. I should like to have formulated analogous principles with regard to nutrition,† but that seemed too wide a subject for the present volume; still, I take this opportunity of saying that

* See Fig. 31, p. 245.
† See chap. xvi., "Analogy between Trophic and Kinetic Actions."

I think the detailed and experimental method of studying movements in the body, and their causes, may throw much light upon the laws and processes of nutrition and evolution. It was a part of my original plan to deal in this volume with some of the laws of nutrition; but, as the work advanced, it seemed better to defer dealing with this subject, as also with regard to some evidence as to the forces at work in producing brain evolution in the individual:* so here we deal with the principles and modes of expression of hidden conditions as a matter preliminary to other studies.

Inasmuch as these studies were primarily undertaken as an attempt to discover the physical signs of the brain conditions which lead to the potentiality for moral, and intellectual character, it is not surprising that they led to the study of the physical signs of the brain conditions giving capacity for "mind."† Still, I was unwilling here to enter upon the physical study of mind, and have dealt only with certain matters preliminary thereto.

In dealing with the phenomena of mind, the present work is a preliminary stepping-stone; if we observe and analyze all physical phenomena coincident with the manifestation of mind, we may obtain data for studying the action of the physical causes aiding the development of mind. This method of procedure appears to me more likely to lead to practical results than any subjective process to introspection of the feelings.

* See chap. xvi. † See chap. xiv

In accordance with these views, I have endeavoured throughout this work to avoid speaking of consciousness, or feeling, as the cause of any mental phenomenon.* It has not, however, been convenient always to do so, but I hope that this rule has been adhered to in all important statements.

Movements and the results of movements, being purely physical actions, are the criteria looked for as indices or expressions of the various brain states. No attempt is made to form an idea of what life, nutrition, mentation, or any other vital property or process may be; the signs of vital phenomena are dealt with, not the living origin of those signs. The term "mentation" is used to imply the function discharged by the brain when the phenomena of mind are displayed.

Frequent reference is made in the succeeding chapters to observations on infants and young children. It appears to me a strictly scientific method to observe and classify the signs which indicate the potentiality for mentation, the early and advancing signs of mentation, and concurrent phenomena. The earliest of these signs are movements and the results of movements, and these are purely physical phenomena, capable of being submitted to experimental inquiry.

Reference is frequently made, by way of illustration, to examples in vegetable life.† If we really accept some form of the evolutionary theory as a working hypothesis, it is reasonable to make

---

* See "Expression of Consciousness," chap. xiii.
† See chaps. ii. and vii.

analogies from the processes proven in plants to processes seen in higher organisms. If physical forces have played a large part in bringing about evolution, we should surely study the processes in simple organisms. If light and gravity are proven to cause certain phenomena in plants, and if we see similar phenomena in men, why should we assume those phenomena in man not to be due to light and gravity, but to be due to "mind," or "feeling," or "consciousness," which we cannot by any scientific process directly deal with?

It has often struck me that recent biological inquirers, in the just desire to study only what is capable of physical investigation, have examined material structures rather to the neglect of the equally materialistic forces displayed in the matter, and that motion which has been so carefully and satisfactorily studied by the physical experimentalist * has not been sufficiently studied by the physiologist. I think that movement in living beings is capable and worthy of detailed study. Following on these lines of thought, this work has been written, showing how largely the expression of vital functions may be described in terms of movement—and movement is capable of physical investigation.

In observing living organisms there are two principal methods that may be employed. (1) Observation of the body,† or corporeal, material structure of the subject: this may be a histological

---

* The works of Grove and Tyndall.
† See specially chap. xvi. as to proportional development.

inquiry. (2) Observation of the ingoings and outcomings. It is the latter set of signs that are principally studied in this volume.

As to the observation of the ingoings and outcomings of the subject observed, the following propositions may be laid down :—

(*a*) The ingoing quantities of material equal the outcoming quantities of material, plus or minus any change in the quantity of material of the body of the subject.

(*b*) The ingoing quantities of force equal the outcoming quantities of force, plus or minus any change in the quantity of force stored in the subject.

(*c*) Some force must be ingoing if any change occurs within.

(*d*) The difference, or change in kind, between the income and the outcome is due to changes occurring in the organism.

(*e*) Conversely, changes occurring in the organism may in part be expressed by describing the difference, or variability in kind, between the income and outcome.

(*f*) It follows that we may reasonably observe and describe the difference between the ingoing and outcoming forces, and also observe changes occurring in the material structure of the subject.

The animus and object of the inquiry described in chap. xix. is to determine, by means of observation and an experimental method, the effects of external circumstances in aiding and producing the physical or material basis of mind.

The primary assumption is made that mentation is dependent upon the material structure of the body, and that the structure, properties, and functions of that body are the result of external forces.

When any principles are laid down for practical guidance in this inquiry,* the endeavour has been made to show that they are widespread "uniformities in the operations of nature," or laws applicable to questions outside the proper scope of this work, and that they are laws or uniformities in the operation of nature of wide action in various biological problems. The endeavour has also been made to show that these laws, as defined and explained, give the foundations for a true experimental inquiry, and that they are in harmony with many of the laws defined in biology, in psychology, and physiology, also in some cases with the laws of sociology and human social life.

It seems to me a matter of great importance to study the attributes of every property observed, the time, quantity, and kind of property; in some cases the "kind" can be described in terms of time and quantity. This is entered upon and illustrated in chap. xvi.

The principles and methods of analysis used (see chap. xv.) are the outcome of daily observations, and have been applied over and over again to many cases in dealing with my living patients. This is mentioned in the hope that others may be induced to see whether these principles are in accord with

* See specially chaps. ii.-iv.

logical and scientific procedure, and as a justification of this attempt to introduce new inquiries into the nerve-system. I have omitted any attempt at cerebral localization, and do not think that the present work has greatly suffered thereby. The work done by others in that direction has been of immense service to me, and possibly this work may indirectly forward that most desirable inquiry. Many purely medical facts might have been introduced, but it appeared best not to produce them in this volume.

The present work has been sent out under its present title as an instalment of a larger work in hand. I think that the accurate observation of the motor outcome of brain action, and the observation of the effects of external forces upon such functions, will give us much information as to the effect of physical forces in causing the evolution of the individual, and the development of mind by educational processes.

Glancing over the succeeding chapters, they may be classified as follows:—

The first five chapters explain the scope and general method of the work. The scope and meaning of the term "expression" is explained to include all outward manifestations of hidden things. Thus it is seen that vital phenomena can only be studied by their expressions, or physical signs; on the contrary, all vital and inscrutable phenomena may be studied by these expressions, if we know how to observe and record the physical signs of the expression.

A certain number of facts in physiology and pathology are referred to in chaps. vi. and vii. It is only by the study of such facts that the significance and explanation of the modes of expression in man can be elucidated, and I hope that the principles defined may be applicable to the description of certain pathological states, especially such as chorea and epilepsy. Pathological anatomy alone does not explain these; probably they will be best described in terms of movements and analyzed by these signs. Direct analogy may be made between the area of nerve-muscular signs in hemiplegia from a destructive lesion of brain, and one-sided chorea, and brain fatigue or excitement affecting the two hemispheres of the brain unequally (see chap. vii.).

The chapter (xv.) on the analysis of expression gives examples of the application of these studies to matters of daily life. It is in particular hoped that this account will be of use to those who study children, and need to read in their outward expression the actual state of their nerve-system.

Artists and those who give literary descriptions of the mental and emotional conditions of man have taught us much, and to them the study of the modes of expression is a necessity (see chap. xvii.). Certain extracts have been given (see chap. xviii.) from some of the older writers, which show that, though modern knowledge may make it more easy to speak with precision of the principles involved in expression, still the matter has long been studied and the thoughts here formulated have been long extant. A bibliography is attached to chapter xviii.

CHAPTER II.

EXPRESSION.

The term "expression" explained—Speech expressive of life; speech a criterion of life—The uniform coexistence of two phenomena makes the one an expression of the other—Abstract properties, and their objective signs; criteria of mind—Comparison of an idiot, and an intelligent man—Certain sounds and a tracing of the pulse express the action of the heart—Signs due to an afferent force; action of the sight of an object upon children; a thermometer indicates heat; the sensitive flame—The telephone exhibits impressionability; a receptive part, an expressive part, and an additional force thrown in—. The phonograph exhibits localized impressionability, which is permanent—Impressionability and retentiveness in a child—Expression, direct and empirical—Impressionability and retentiveness—Nutrition; its signs—Movements in plants; pulvini, by unequal growth—Expression by form, colour, temperature—Vital processes can only be studied by their expression; the importance of appreciating this in biological work—Summary.

"EXPRESSION," in its widest signification, is the outward indication of some inherent property or function. An expression is a physical sign which is accepted as a criterion of the property, because the two are found by experience to be more or less uniformly coexisting phenomena; and if the manifestation of the special expression (or physical sign) is found with absolute uniformity, whenever the property or function in question is present, we may look upon that physical sign as an absolute proof of the presence of the property. Thus, in a man the power of speech proves the existence of "life;" physiological knowledge shows that we cannot have speech proceeding from a man unless his body has life. To prove the property life in a human subject it is quite sufficient to prove that he can speak. We do not know what life is in the abstract, but to prove that a man can speak is to prove that he possesses what we call "life."

It follows, then, that the physical sign "speech" is a criterion of life, or we may say that speech is one expression of life. We see, then, that the uniform coexistence of two phenomena in a subject, of which one is a physical sign, enables us to look upon that physical sign as an expression of the other phenomenon.

This uniform coexistence may be demonstrated by empirical experience sufficiently extended, and attested by various observers, under varying circumstances. If we analyze the expression of the property "life," given above, we shall learn something further as to why we may speak of "speech" as an expression of life in the subject. It is known that speech is always the result of certain movements of the respiratory apparatus, the tongue, and the mouth. Respiration, even when unattended by speech, and when found in animals incapable of speech, is uniformly expressive of the property "life." Further, physiology has shown that active properties and functions of the brain are necessary for speech. It follows then, that speech is an expression of two phenomena in the subject—respiration and brain action.

Speech is an expression of respiration and brain action, because it results therefrom, or is an outcome of respiration plus brain action; so this illustrates the truth that, if a certain physical sign is always the result, or outcome, of some one property or function, that physical sign is an expression of that property or function. It is often convenient to study at the same time a certain property in the abstract, and its outward or physical signs. When we study "mind" as a property of man, we must define its physical signs, or the criteria by which we appreciate the function "mind." If we compare a healthy and intelligent man with an idiot not possessed of the properties called "mind," or possessing them only in a very low degree, we shall soon see a marked difference in the physical signs. The intelligent man speaks well, his attention is attracted by objects of beauty or usefulness. The idiot does not speak intelligibly, his attention is not attracted by objects of beauty, his movements are not subservient to his own wants. A man of mind affords expression by the manner in which his attention is attracted, by his good speech, and by his movements; these objective facts are, then, the expression of his mind. Any property that can be possessed by a subject may be indicated by some physical sign, which is then called the expression of that property in that particular subject. Temperature is an expression of heat in the subject; growth in the seed is an expression of life in its organism.

We have briefly touched upon some forms of expression of the brain, which is an organ of the body, the special forms of expression considered being the outcome of its properties and functions. We will now discuss the expression of one or two other organs. We form a judgment of the condition and functions of the heart by the state of the pulse, the sounds heard over the heart on auscultation with the stethoscope, and by feeling the impulse of the heart, etc.; to form a more accurate judgment we may take a tracing on paper of the movements of the pulse by an instrument termed a sphygmograph, and we obtain a line on paper, indicating the movements of the heart, by means of another instrument called the cardiograph. The sounds and the characters of the impulse of the heart, and the tracings of the movements of the heart and pulse, are expressions of the action of the hidden organ—they are the outcome of its action, and the criteria by which we judge of its functions.

We have hitherto spoken of modes of expression where certain signs indicate properties, either because they are found uniformly to coexist with those properties, or else because they are the direct outcome of such properties. We now come to consider how expression may be a result of some force afferent to the subject, falling upon it, but not solely and directly the outcome of its intrinsic properties and functions. This may be conveniently termed "expression by reflective action."

It is a matter of common observation that we do not know a man's mental and moral qualities till we have seen him tried by the circumstances of life. In forming a judgment of a man's character we seek to observe his actions under various circumstances. The man's actions under these various circumstances are partly the outcome of his organization and character, and are partly due to the special circumstances acting upon him. Such expressions are in part the effect of circumstances acting on the man. Such a case illustrates this subject, but is too complex for a convenient analysis.

Let us take a simple example. Suppose two children, both healthy, intelligent, and of the same age. Hold a pleasing toy before one child, and let it be hidden from the other. We now see the expression of pleasure in one child's face, accompanied by gestures of delight; not so in the other child; but if both are then allowed to see the toy, each will present expressions of joy. The sight of the toy is the cause of the expression of joy in the children. Here, then, the expression is not solely the manifestation of the spontaneous outcome of the organism of the child; the sight of the toy falling upon the children is necessary to stimulate the expression of joy or amusement.

Again, if an experiment be tried by placing the toy in front of the child while he is asleep, or if awake when very ill, no expression is produced; or lastly, if the child be very cross and sulky, the sight of the toy may cause no expression of joy. If, on the contrary, the child be "in a very laughing humour," the expression by similar stimulation will be excessive in degree.

These considerations show that the condition of the subject is in part indicated in expression when the subject is acted on by external forces afferent to it. In speaking of reflective action as a mode of expression, it will be seen that the line of argument founded upon such an observation may be inverted. If the subject is known to be constant and unchangeable in its properties under the same conditions, any variation in the outcome of its properties or functions is due to a change in the afferent forces or the environment. In looking at a man of known stable constitution, in whom no disease or special defect of any organ is discoverable, if we observe some sudden change in the aspect of his face indicating, according to our experience, grave mental anxiety, we may conclude that something has happened to him causing anxiety—some news of disaster or cause of fear, etc. The aspect of his face is the expression, firstly, of his condition; secondly, it affords evidence that some event has happened affecting him with fear.

A thermometer is an instrument not changing itself; any variation in the volume of mercury in the bulb and stem is due to heat: here any increase in the bulk of the mercury is due to heat coming to the instrument, and the rise of mercury in the stem expresses the temperature of the surrounding medium.

The sensitive flame is another example of a subject so impressionable that it indicates or expresses very faint sounds or aerial vibrations. The following quotation is from a lecture by Mr. W. F. Barrett, delivered before the Royal Dublin Society, January 8, 1868:—" The lecturer had reserved for the conclusion a flame wonderfully sensitive to the slightest noise. The burner which gave this flame was formed of steatite, and consisted of a single circular orifice, through which the gas was forced from a large holder in the lecture-room, with greater pressure than would be attained from the main. The flame was now fully two feet in length; 'and observe,' said the lecturer, 'how delicate and fragile a thing it appears to be; for on the slightest noise it drops down a foot. The jingling of this bunch of keys, the crumpling of this paper, the dropping of a small coin, are more than sufficient utterly to break up its height and symmetry. This flame makes no response to the vowels *o*, *u*, nor to the labials, but it energetically responds to the sibilants. Repeating the stanza—

> " Roll on, O rill, for ever!
> Rest not, lest thy wavelets,
> Sheen as shining silver,
> Shrink and sink to darkness;"

the flame is unmoved by the first line, but emphatically bobs at the sounds "rest" and "lest," and admirably suits its action to the words of the last line, for, when shrinking, the light of the flame almost disappears. So sensitive is this flame that even a chirp made at the far end of the room brings it down more than a foot. Like a living being, the flame trembles and cowers down at a hiss; it crouches and shivers as if in agony at the crisping of this metal foil, though the sound is so faint as scarcely to be heard; it dances in tune to the waltz played by this musical box; and, finally, it beats time to the ticking of my watch. How wonderful are all these facts! And the more we know of them the more wonderful do they appear; for this astonishing change in the aspect of the flame is produced by an infinitesimal portion of those almost inaudible sound-waves, already enfeebled by their distance from the flame.'"

In the telephone we see an instrument of great impressionability, but it exhibits more than this. Two instruments being placed in electrical communication, and the instrument having a battery in the circuit, the vibrations of the voice communicated to one instrument are conducted to the other; that is to say, the one instrument is impressionable, the second expresses the vibrations received by the first, while the force received from the battery is an active agent or factor in producing the expression. In this arrangement we see an impressionable part, a conductor of impressions, an additional force thrown in, and a part solely utilized in producing expression.

The vibrations of the receiving disc are ended as soon as they arrive, and no permanent impression is left; there is no retentiveness of the impressions in any part of the apparatus, no permanent impressionability.

In the phonograph we find impressionability more or less permanent; but, while receiving impressions, the instrument cannot give origin to expressions. The one vibrating plate receives the impressions, and subsequently expresses them, retentiveness being concentrated in the tinfoil. As in the telephone, so in the phonograph, an additional force is thrown in. While the impressions are being received by the sounding-plate and impressed upon the foil, it is, of course, necessary that the foil be moved under the needle that indents it, in order that fresh parts of the foil may receive the indentations. The impressionability in the arrangement may be said to be permanent, the indented foil remains indented for an indefinite length of time. We best know the property retentiveness in the apparatus by the expression of the sounding-plate when the cylinder is again moved so as to reproduce the original sounds which came to the sounding-plate. The retentiveness is indicated by the efferent expression obtained as sound on causing the foil to revolve; the expression corresponds more or less exactly to the afferent sound producing the impression, although a long period of time may have elapsed between receiving the impression and giving out the expression. Another thing to be remarked upon as essential to getting expression out of the phonograph is that the indented foil must be made to move the sounding-plate, and mechanical force must be applied to the cylinder carrying the foil, so as to cause this to move under the needle of the sounding plate, making it vibrate.

Impressionability and retentiveness may be observed in animals acting in modes almost analogous to those seen in the telephone and the phonograph. Take as an example a child whose attention is attracted by a musical box in action, so that it turns towards it and listens. The sound of the musical box causes vibrations in the child's apparatus of hearing. Suppose that the child is playing with a toy when the music commences; attracted by the sound, the child's head turns away from its toy towards the musical box, and he claps his hands. This is one effect, one expression of the impression received. Further, the child will, after a certain number of repetitions of the same tune, be so far impressed as to be able to hum the tune, in part at least, under the stimulus of pleasure or a request from the mother; this indicates the permanency of the impression. The action of vital force in the child is, of course, necessary to reproduce the tune; this corresponds to the mechanical energy supplied from without to the cylinder of the phonograph.

Another class of expression that must now be considered is what may be termed "empirical expression." Here the expression, or objective sign observed, is not the direct outcome of the intrinsic property or function which it indicates, but we infer from the presence of the objective sign that the special property is present. This inference is founded on the observation of the uniform occurrence of the objective sign and the intrinsic property, although the two are not related as cause and effect, as in cases of " direct expression." Of course, it may be argued that in this empirical expression the intrinsic property, and the observed objective sign, are both results of a common antecedent cause; thus both may be inherited alike. It is a matter of great importance to our subject that the difference, and relative value, of empirical and direct expression should be thoroughly understood and appreciated, because it is only the modes of direct expression that can be looked upon as direct evidence of the possession of certain properties.

When we look at a living human face totally devoid of movement, or special expression as the result of movement, or even if we look at a plaster cast of the face, it may indicate to us something of the character of the subject. In looking at the impassive face, and the plaster cast, the expression seen is not the direct outcome of brain action, or the condition of the mind. We are speaking here of the more permanent and fixed conditions of the expression, not of fleeting, transient, mobile conditions. Still, there is the fact that the two things—the average action of the brain in marking the average of the man's mind, and the form and contour of the face—coincide empirically; this subject will be more fully considered further on;* it is the foundation of the scientific study of "Physiognomy." Examples of

* See chap. xvi.

empirical, or indirect expression, may be seen in organic, and in non-living subjects. When such cases occur in man, or in living animals, or vegetables, they may be termed "examples of coincident development." This empirical expression differs from the modes of direct expression, inasmuch as in the latter, the expression is the direct outcome of the property expressed. The sprouting beard in a boy's chin, is a sign of commencing manhood, not because the one produces the other, but because a widely extended experience of many boys has shown the two things to be usually coincident. Pink-flowered geraniums are plants of feeble constitution, not because the colour of the flower directly affects the constitution of the plant, but because pink-flowered geraniums usually have highly coloured leaves, with but few cells containing chlorophyll, and the presence of chlorophyll is essential to the nutrition of the plant. Pink flowers are, then, only an empirical sign of feeble constitution in the geranium, but still a sign of value; the real or direct expression of the constitutional feebleness is the small amount of chlorophyll in the leaves—the scarcity of chlorophyll is the direct expression of the feeble constitution, because it is the cause of it.

Impressionability and retentiveness are properties found alike in living, and non-living subjects; nutrition and growth belong to living beings only, or so it is generally considered. By the term "nutrition" is denoted an active vital process, converting pabulum into tissue of the organism, or into some new chemical material, thus evolving force which may be observed in action, either at the point where it is produced or at a distance from it.

Nutrition being a vital, and inscrutable process, we probably cannot know it in the abstract, but we can observe the effects of the process of nutrition in an organism. The best evidence of the occurrence of nutrition is found in its results. In many cases nutrition is greatly aided by (if not entirely dependent upon) the action of certain external forces afferent to the subject, so that the results in some degree indicate the action of such external forces. We will consider a few examples. In vegetable growth we may find a simple multiplication of cells, all similar in their histological or structural characters, as is seen in the tissue termed "primary meristem." Here the expression of nutrition, and vital action, is multiplication of cells. In a mass of unicellular plants the result, or expression of, nutrition is multiplication of plants, as in the growth of the snow-plant. In other cases, as in the apex of the stem of a plant, when a bud is being formed, differentiation of the cell-growth results from nutrition, producing embryonic leaves. A more complicated case of nutrition expressed by histological changes, is seen in the cellular body called an ovule, which is found in the carpel of a flower. When the protoplasmic contents of a pollen grain have entered that cellular body, changes follow which cause an embryo to develop by cell-differentiation in the ovule; the structural changes observed in the development of the embryo are the expression of the vital effects of fertilization, or sexual impregnation. Other cases of vital action in plants are differently expressive. The action of chlorophyll in the cells of a leaf is known by certain chemical results—carbonic acid gas being absorbed, the oxygen being discharged from the plant, and the carbon retained in its tissues. Here the chemical changes are the expression of the action of chlorophyll. The growth and vital action of the yeast-plant, when living in a solution of sugar, is in part expressed by multiplication of the unicellular yeast-plant—a histological mode of expression—and it is in part shown by the chemical changes of fermentation indicated by chemical results.

Numerous kinds of movements occur in plants; two examples will serve our present purpose, showing that movements may be the expression of vital changes. In plants, movement is effected by two principal methods—by unequal growth of the cells composing the growing member; or by organs termed "pulvini," temporarily or permanently devoted to the production of movement as their principal function. In the *Oxalis* (wood-sorrel tribe), and in the sensitive plant (*Mimosa*), movements of leaves are seen. In each case the movements are effected by an arrangement of cells at the junction of the leaf with its main stalk; this group of cells is called a "pulvinus." The pulvinus is the mechanism by which movement of the leaf is effected; it consists of a mass of small cells destitute of chlorophyll, and therefore incapable of performing any nutritive function in the plant, or of taking any direct part in the elaboration of its nourishment. This pulvinus is the lower portion of the petiole, or leaf-stalk; and the movements of the leaf depend upon its cells, which expand alternately, first on one side, then on the other. Structurally, the pulvinus consists of small cells arrested in their development while still young. These,

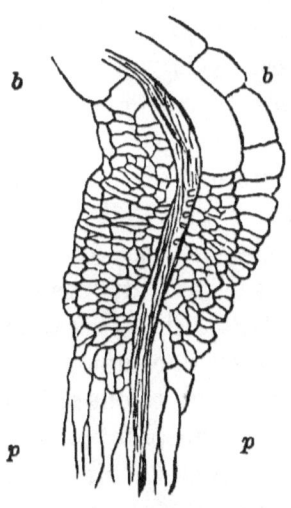

Fig. 1.—After Darwin. Longitudinal section of a Pulvinus, magnified seventy-five times. *p p*, petiole or leaf-stalk; *f*, fibro-vascular bundle; *b b*, commencement of blade of cotyledon.

when turgescent with sap, swell up, thus increasing suddenly the bulk of the structure composed of them: and the turgescence leads to motion only—it is not followed by growth; no nutrition of the plant results from the turgescence. It must be noted that the cells possessing this special function of producing motion only, fulfilling no direct nutritive purpose, are smaller than their neighbours, arrested in growth, and destitute of chlorophyll. This is the condition of the cells of the pulvinus as long as it is capable of producing movement. In the *Oxalis corniculata* the pulvinus is developed imperfectly, and to an extremely variable degree, so that it is apparently tending towards abortion. Its cells contain chlorophyll; that proves they have nutritive functions to perform, as well as the production of movement. These cells are more like the normal cells of the petiole than those usually characteristic of a pulvinus.*

As to the production of movement in plants by unequal growth of the cells. In the process of germination of the pea seed, the radicle of the embryo protrudes from its case, and direct experiment demonstrates that the apex of the root in its downward progress does not proceed in a straight line; the apex bends first to one side, then to another, moving in irregular ellipses, or rather in a spiral direction, as it descends. This zigzag movement of the apex of the root is due to an unequal growth of the cells composing the growing portion of the root; they enlarge first on one side, thus causing the apex to bend towards the opposite direction, and then the set of cells growing alters. This mode of growth is termed by botanists "circumnutation." The essential principle of this mode of growth is, that the vegetable cells concerned do not grow together, but those on one side grow, then those at another

* "Analogy between the Movements of Plants and the Muscular Movements of Children, called Chorea," F. W., *British Medical Journal*, February 25, 1882.

part, in regular series. In this circumnutation movement results from the fact that the cells grow in series, not all simultaneously.

Form, colour, temperature, are properties as frequently observed in non-living as in living subjects. These may, perhaps, be looked upon, and spoken of most conveniently, as obvious properties because the form, colour, and temperature are their own expressions. This statement is not exactly true, but this form of language does not convey any untruth that will vitiate our use of this term for present purposes. Strictly speaking, "the colour of an object" is a condition of its surface, and what we see and describe, is the reflection of light from it, and the kind of light so reflected is an expression of the molecular structure of its surface; "the temperature of a body" is an expression of the condition of the vibrations of its molecules; "the form of an object" is the result of the forces that produced it.

The largest possible scope has been given in this chapter to the term "expression," and this has been done with a definite object. Expression is the objective sign of a property of the subject. In biological inquiries we are obliged to study vital processes by their expression in objective signs. Life itself is studied by the results, or objective expressions of nutrition, and growth. It is, then, obviously desirable that we should study all the objective signs, or modes of expression of these vital processes. The subject of expression is dwelt upon in the wide signification here given to the term, because it appears doubtful if, in biological research, dependence enough has always been placed upon the truth of the fact that expressions are objective signs of the properties of the subject. If we are certain that the objective signs seen in expression are either uniformly accompanied by certain phenomena, or that the cause of the special hidden phenomenon is the reason of the visible objective sign,—in either case we may study the hidden condition by studying the outward objective expression thereof.

Thus, "mind" in the abstract cannot be submitted to direct physical investigation by experiment, but, if certain physical signs can be shown to be the criteria or expression of mind, we may be able to observe these objective signs, and devise experiments for the elucidation of information with regard to their cause. Again, nutrition of a living organism is a subject of which we understand but little, and what we do know is the result of studying the objective effects of nutrition, and its coincidences—in other words, the objective expression of nutrition.

Summary.—In this chapter I have endeavoured to explain what is meant in this work by the term "expression," and it is shown that the scope of our subject is wide, dealing with any case where objective physical signs indicate some more hidden phenomenon or condition. When a physical sign is a criterion of a condition of the subject, we may investigate that condition by studying the physical sign.

The physical sign or expression may be the

direct outcome of the condition it indicates, as speech is the direct outcome of respiration and brain action; or the expression and the inherent condition may be found by empirical experience to be uniformly coincident. The expression, or outcome from the subject observed, may be the result of some force afferent to the subject, as the movement of a sensitive flame, which is caused by vibrations of the air,—there the physical sign observed, the movement of the flame, is an expression of the vibrations of the air. The physical signs observed may express impressionability, temporary or permanent; in the telephone the receiving plate is very impressionable, but the effect of the vibrations it receives is temporary; in the phonograph the vibrations received are impressed upon tinfoil, and more or less permanently retained. Impressionability, temporary or permanent, is frequently expressed in plants, animals, and in man.

Those cases may be termed "direct expression" where the physical signs observed are the direct outcome of the property expressed; an "empirical expression" signifies that the physical sign observed is uniformly found by experience to be associated with a certain property, though the line of causation may not be understood. Nutrition is a property belonging only to living beings; the term denotes an active vital process converting pabulum into tissue, or evolving force which may be observed in action. The results of nutrition may be in part due to external forces afferent to the subject, and so may in part express them.

Form, colour, temperature, are properties that may be expressive in living and in inanimate objects.

The movements that occur in the growth of plants are expressive of their modes of growth.

CHAPTER III.

EXPRESSION IN MAN AND IN ANIMALS.

The act of expression does not prove vitality—In living things expression is the outcome of nutrition—Nutrition not considered here, only its expression—Expression when the outcome of processes in the subject is called direct—Growth results from local nutrition; such processes are termed trophic—Nutrition is an expression of life—Permanent impressionability; it is opposed to evolution, it may be expressed by reflex action—Retentiveness need not be permanent.—Development and reflex action as modes of expression—Apparatus for reflex movement; reflexes may be congenital, or acquired—Reflected action, as expression by form or colour—Any outcome of function may be expressive—Expression by colour, sound, change of function in a part—Movements of an actor in anger—Trophic action illustrated by the growth of crystals, growth of the body—Coincident development of parts; coincident defects in imbeciles—Properties demonstrated by external forces—Heredity—Expression of the emotions—Summary.

IN the last chapter the principles and modes of expression were described and discussed, and it seemed well to illustrate all points by examples taken from living and non-living subjects. The mere fact of expression in a subject is no evidence of vital characters, no proof that it is a living rather than a non-living thing. In the present chapter we shall speak of expression in man and in animals, and the illustrations of modes of expression will, as far as possible, be taken from the cases of living beings. The great and striking general difference between the modes of expression in animate and inanimate subjects, is due to the fact that, in the former, vital properties and nutrition are essential.

We will commence with the study of the signs of nutrition. The differences between the modes of expression in animate and inanimate beings is, then, due to the fact that, in the former, vital properties and nutrition are essential factors. Expression in the living subject is usually the direct objective outcome of conditions of nutrition; we therefore consider together the subject of nutrition and the expression of nutrition. As to nutrition *per se* we have very little to say here, but we are deeply concerned with the modes in which nutrition is expressed. It has been said that the most certain modes of expression are those which are the direct outcome of action in the parts involved; therefore the best modes of expression are the objective results of nutrition. Now, the results of nutrition being many, it will serve our purpose to deal with a few cases only—growth, movement, evolution, retentiveness. Growth is an objective sign of nutrition—a corporeal, structural, or histological change in the subject; the material structure of the subject, not merely its properties or functions, being changed as the result or outcome of this form of nutrition. Local nutrition is necessary to growth, therefore growth is a sign of nutrition. Rapidly growing tissue is believed to be highly nourished, because it is found to be very vascular; and often we may infer the degree of nutrition from the degree of vascularity.

That kind of nutrition or vital action which produces growth, or change in the material structure of the subject is conveniently termed " trophic action ;"; we may speak of expression by " trophic action," and compare it with expression by movement. When trophic action is the mode of expression, we find some histological change in the subject of growth; when a movement is the mode of expression, there is often no change in the structure of the subject that can be observed as a histological fact, and if we neglect to observe and record that movement, we possibly neglect the only objective sign of the change occurring in the subject. Nutrition and growth, when they occur together, indicate or express life in the subject; it follows, then, that any of the expressions of nutrition and growth are expressions of life.

Permanent impressionability is a very important property, and it was shown in the second chapter that retentiveness may alike be found in animate and in inanimate subjects. Retentiveness in living things may be indicated or expressed in various ways. When in a subject a certain stimulus is always followed by the same expression, we may conclude that the subject is unchanged, or retentive at least of the property thus expressed. Retentiveness is, then, not a sign of evolution or change, but of an organism capable of resisting changes. Retentiveness indicated by reflex action may be seen in the infant at birth. The fact of placing an object between its lips excites the movements of sucking; this is a reflex action, and persists as long as the individual is an infant. When the infant becomes a man, this reflex is more or less weakened or lost; the mechanism connected with sucking is not retained in the same condition through life. We see, then, that in this particular mechanism retentiveness is only temporary.

Movement, as an outcome of action in an animal, is the result of some force, internal (as nutrition) or external, acting upon the animal. The study of the correlation of forces justifies this assumption that visible movement, as the outcome of an organism, expresses either force afferent to the subject, or the result of changes occurring in it—changes which are conveniently termed vital action or nutrition. As examples showing movements expressive of nutrition in the organism, the following may be cited. Mechanical exercise in man, such as carrying a weight, has been shown by physiologists to be due to changes in the body, and the movements of the man are an expression of the internal changes. A child in perfect physical health is frisky, plays, runs about, and chatters incessantly till he is tired; the movements indicate the perfect physical health of the subject, because they are the outcome of the perfect nutrition of the organism. If the nutrition of the body at large is low, as the result of deficient or improper food. or disease of some organ, the lowered condition of nutrition is indicated by the lessened amount of movement. The dormouse when hybernating has its nutrition lessened, owing to the little food it takes; its feeble circulation, its lowered temperature, and the absence of movements in the limbs are the expressions of its lessened nerve-muscular energy.

Development is a very interesting and important study. It is expressed in various modes. In the newly born infant we observe certain conditions indicated by its weight, form, proportions, etc., *i.e.* certain conditions of its body; we also observe certain functions, movements, reflex actions, etc. These are the expressions of its condition. In the adult the weight, form, proportions, and the movements and reflexes, are different from those of the infant, and this difference is an expression of the evolution of the individual. After comparing many infants with many adults, and observing the development of infants into adults, we find that a more or less regular series of similar changes occurs in the body, and in its functions, as development advances from infancy to adult age. This more or less regular series of changes in the body, and in its functions, is the expression of the development, and, in any case, before we can give a full historical account of the development, we must observe all modes of its expression. The life-history of an individual animal is in part the history of its development; it includes the description of the body of the individual at every period of its existence, together with all the influences afferent to it, and all the expressions, or efferent functions, coming from it.

Reflex action is a mode of expression of common occurrence in man and animals; this will be more particularly dwelt upon as a mode of expression in the next chapter. Drawing a hand away from a hot cup is a reflex action; but the movement is a mode of expression, such movement being usually considered expressive of pain. For the occurrence of a reflex action a certain nerve-motor apparatus is necessary. The simplest apparatus for this purpose consists of (*a*) sensory surface; (*b*) afferent nerve-fibres; (*c*) nerve-cell or centre; (*d*) efferent or motor nerve-fibres; (*e*) muscle. Now, this apparatus for a reflex action may exist at birth, as with the apparatus for sucking; or it may be acquired, as the reflex movement of the face giving expression of pleasure at the sight of a fine work of art,—such reflex, not existing at birth, is said to be acquired. We may, then, have expression by reflex action congenital, or acquired. It seems also probable that reflex actions form an important and very large share of the phenomena grouped as the function "mentation," or the faculty of the brain which produces mind.

Reflected action is a mode of expression in living and non-living beings. The term "reflected action" signifies a purely passive condition of the subject, all action being on the part of some force afferent to the subject. The expression of a statue is purely passive—the marble does nothing itself; the rays of light falling upon it are in part reflected, enabling us to see the statue. The expression of the passive features in man is reflected action; so all expression by colour and pigmentation, form and proportions. Reflected action is not *per se* an indication of vitality or nutrition. When in an animal movement occurs as the result of some mechanical irritation, as tickling, the afferent mechanical agency produces a change upon the sensory surface which causes a stimulus to pass to the nerve-centre, from which an efferent motor-current then proceeds to the muscles, producing movement. Such movement is called reflex action, or reflex movement, in distinction from the case of the statue, where there is no change or movement in the subject, which is passive, all expression being an offcome, not an "outcome;" the objective cause of expression is the reflected afferent force.

Expression may be effected by colour, or any other mode of objective outcome or offcome passing from the subject of expression to the observer. Conditions of colour are often expressive. The colour of minerals often expresses their properties; the colour of parts of plants is often very constant and uniform, enabling this character to be used in some degree, as a sign by which an individual may be recognized and classed. The special hereditary characters and breed of many animals are often expressed by conditions of coloration, striping, spots, etc. Light complexion of face, light hair, etc., are expressive of race and climate, equally with dark pigmentation and olive complexion. In point of degree of value as signs, there is no doubt that structural points are of more importance than points of colour. The colour is the result of the molecular condition of the surface; the general structure is a more widely spread physical condition.

Sound is one of the commonest modes of expression among animals. The sounds produced are usually indirect expressions.

It is very interesting to study changes in the functions of an organism or in part of a living being, and such studies are very important in any investigation of the processes of development. We have, then, to look to the signs or expressions of change of function in an organism, and the most satisfactory kind of evidence is that afforded by "direct expression," or some change in the outcome of the action of the subject which can be directly observed as an objective sign. A change of function in parts of the body of a plant or animal is not an uncommon occurrence, and may serve as a means of expression. In the life-history and development of many species a change of function is uniformly met with. A few examples may be mentioned, taking together the change of function and its expression in objective signs.

The case of change in function in the cells of the pulvinus of *Oxalis* has been already given in chap. ii. p. 26.

The body is not fully developed in childhood, and the hair-bulbs in the skin of the face take on active growth about the time of the development of full manhood. In a woman the growth of hair-bulbs in the face is usually delayed till old age. The functions of the stomach, and its power of digestion, vary at different periods of life, and under different circumstances: the stimulus of food causes an acid secretion to be poured out; when empty, the secretion of the organ is alkaline. In the fœtus the lungs are not used as respiratory organs, and at birth the salivary glands are not active. Ants, when first born, possess wings which they soon learn to use, but after a short time they lose their wings and become solely terrestrial animals for the rest of their days. A bar of iron may be magnetized, and its properties are thus changed. In vegetable growth it is very common to find examples of epinasty and hyponasty* alternating, the cells growing first on one side, then on the other.

In man certain changes in the larynx in the course of growth cause an alteration in the voice. The action of glands varies from time to time; *e.g.* secretion of mucous membranes, etc. If a man is kept long in the dark his retina becomes oversensitive to light. Definite and permanent changes in the structure and functions of a part are more clearly seen in a species than in an individual; this is owing to the greater length of life of a species as compared with an individual.

Movements are often exceedingly expressive. When observing a man, for the purpose of forming a judgment as to his mental condition, and determining what emotion is most active within him, we take note of his gestures or movements. An actor

* Sachs, "Text-book of Botany," 1875. p. 767.

upon the stage, desiring to represent the emotion anger, imitates, according to his art, the gestures of a man angry in earnest—his features are distorted, he makes movements as if he would combat his adversary, etc. These are the actor's imitative expressions of anger. A child, when angry, stamps, gesticulates with his arms, and cries,—these are real objective expressions of his anger; so also is sorrow expressed by the downward drawing of the angles of the mouth. In chaps. viii. and ix. many examples are given of expression of the emotions by movements as objective signs. The voice, its tone, and its rapidity, are highly expressive of emotions; the voice is the result of nerve-muscular action. A stooping attitude and spiritless gait indicate that a man is tired or dejected, as compared with his postures when refreshed and energetic. The head is drooped as an accompaniment of shame; it is held erect and firm when defiance is expressed.

Physical conditions termed "trophic changes" are often very characteristic and expressive of development, and of nutrition, in the subject. As previously explained (see p. 33), by the term "trophic" action, or change, it is intended to indicate some change in the material structure of the subject, the histological or structural change being itself the expression of what has taken place. Taking one example outside the proper subject of this chapter, the case may be cited of the so-called growth and repair of crystals.* "If a portion of a crystal be

---

See Sir J. Paget's paper, "An Address on Elemental Pathology," *British Medical Journal*, October 16, 1880.

broken off or filed or dissolved away, and if then the mutilated crystal be replaced in a solution of the same salt or of an isomorphous one, the lost part will be replaced, the crystal will be enlarged, new crystallized matter will be formed on every surface; but the quantity formed on the injured part will be greater than that formed at any other part, and repair will be more active than mere growth till the proper form of the crystal is regained. Then, when the repair is complete, growth alone will go on, and each part of the crystal, if it remain in the same solution, will increase in due proportion with the rest." Here the reformation of the injured part of the crystal is the objective expression of the action of the forces at work in producing the perfect and enlarged crystal.

In a man we judge of the general condition of his nutrition, by the weight of his body in proportion to his height, and his general proportions or build in relation to his age. With slight alterations of the amount of general nutrition and health, we see slight alterations in the form and outline of the face and figure. The growth of the teeth is indicative of the age of the animal; grey hair indicates senility or depressed vitality; the size of a fruit indicates the effect of the forces that produced it (see Darwin's " Variation under Domestication "); a slight degree of absorption of fat in the body, as the result of slight changes in the conditions of nutrition, causes a dull appearance of the countenance; the smell of some animals indicates the condition of their general health.

Another important mode of expression in man and animals is that which may be conveniently termed "the coincident development of parts." Observation often shows that two parts of the body usually agree, in having the same proportion of good or ill development. If the one part is well developed, the other part is likewise well developed, and *vice versâ*, although the one part is not the origin of the other, or directly connected with its formation. It may probably be shown hereafter that the good or incomplete development of each of the parts thus found to correspond, is the result of some common antecedent or cause, or that both are alike inherited. Thus we see the colour of the hair, and the colour of the iris, often bear a fixed relation to one another; the height of the body and its properties as indicated by measurement, usually show a marked degree of correspondence in the same race of men, and in animals of the same species if living under similar circumstances. The subject is dwelt upon in the "Origin of Species," p. 115.

Certain properties of a man or of an animal, are only expressed when external forces demonstrate their existence. The capacity to feel joy is proven by the expression of joy when something happens to the subject producing such emotion. The capacity of colour-sense is demonstrated only by aid of coloured objects. In these cases circumstances external to the individual help to express the properties in question.

The property or function heredity is one that must be expressed outside the individual subject. A few words only can be said here on the very wide and deep subject of expression by heredity, that is, expression of the conditions of the life-history of the individual, as demonstrated in its offspring, the offspring showing the effect, or outcome, of forces afferent to one or both parents.

For our present purpose I think we may classify the phenomena of heredity in terms of certain properties: (1) trophic phenomena; (2) motor, or kinetic, phenomena; (3) reflex action; (4) retentiveness. We are here dealing solely with the criteria or objective signs of heredity, not with its essential nature. A general review and consideration of the facts of heredity enables us to say that we may look upon the four criteria, above mentioned, as the main modes of expression of heredity as a property in the subject. This mode of expression will be discussed in chap. xvi.

In concluding this chapter we must refer to "expression" in the more limited sense in which the term is commonly used as regards man and animals. "Expression" is a term commonly used to signify the modes in which we judge from outward appearances of the mental or physical condition of the individual at the time of observation; thus we speak of the expression of pain, joy, intelligence, hunger, sleepiness, etc. It is our business here to study these expressions, these outcomes or ejecta, these uniform objective concomitants of the hidden conditions intelligence, hunger, sleepiness, consciousness, feebleness. It is not for us here to study subjective conditions or feelings and states of consciousness, but we are to study their expression or objective concomitants. All that can be observed are the ejecta, or outward visible signs; the coexistence of the subjective condition is an inference dependent upon the uniform experience that such and such outward or objective sign is always, or almost uniformly, accompanied by a certain subjective condition or feeling. In the expression of physical pain the angles of the mouth are depressed. This statement is justified by the frequent, almost uniform, observation that, when the angles of the mouth are markedly depressed, inquiry shows that there is some source of physical pain in the man or animal; and conversely, that, in many cases where there is known to be pain, examination has shown that the angles of the mouth are drawn down. It has been said by some that pain must be the cause of the depression of the angles of the mouth. We do not know what pain is; it is a subjective condition. We do know that the depression of the angles of the mouth is due to muscular contraction, and that the muscular contraction is due to a nerve-current from the nerve-centre. It is, then, the condition of the nerve-centre corresponding to the muscles which depress the angles of the mouth, that is so uniformly affected when there is a source of pain in the subject; in fact, all the special irritations which cause pain, seem to affect the nerve-centres of the depressor muscles of the angles of the mouth. This is the knowable objective fact.

It seems to me that the ground is cleared for us

in this inquiry by omitting all consideration of the subjective conditions, and considering only the objective facts. In animals the subjective condition is but a very crude inference from the objective facts.

Summary.—The term "expression," as used in this work, does not imply that the subject of which it is connoted has the property life. Nutrition is said to occur only in living beings. We do not know much about the vital process nutrition, but we can observe and study its results, or objective signs. The outcome of nutrition affords the best examples of direct expression. The principal results or expressions of nutrition are—growth, or trophic action; movement, or kinetic action; evolution, and retentiveness.

Growth concerns the material structure of the subject—the kind of action which produces a change in the structure of the subject; it is, therefore, called a "trophic action," in contradistinction to that result of nutrition which only produces movement, and is called in this work "kinetic action."

Permanent impressionability, or retentiveness in a living structure, may be expressed by a reflex action; it is not a process of evolution, but gives a tendency to resist change.

Movement is an outcome of changes occurring in a living subject; this is a deduction from the law of conservation of energy.

Development is expressed in various modes—by the ratio of growth, weight, and proportions (trophic action), or by series of movements and reflex actions (kinetic action). The mechanism for a reflex movement requires (*a*) sensory surface; (*b*) an afferent nerve-fibre; (*c*) a nerve-centre; (*d*) an efferent or motor nerve; (*e*) a muscle;—such apparatus may be congenital or acquired. Probably reflex action has a large share in that faculty of the brain which produces mind, and which is here called "mentation."

"Reflected action" is a term used to indicate a passive condition of the subject, expression resulting from the mode in which it reflects an afferent force, such as light; such is the mode of expression in a statue. Expressions by form and colour are similar modes.

Sound produced by a living being is a result of movement, and as such is highly expressive.

Change of function is an important mode of expression; it may be studied in vegetable cells, glands, etc.

Movements (kinesis) and the results of movement are among the most important modes of expression, and these are conveniently termed "kinetic," in contrast to "trophic modes" which affect the material structure of the subject.

Coincident development, proportional and similar development in the members or parts of a living subject, are interesting modes of expression; they are trophic in kind. These are discussed in chap. xvi.

The facts of heredity are very important and complex modes of expression; some facts concerning a man or an animal are only known by observing the parents and the offspring.

As to the subjective conditions in man, no account can be given of their essential character; we can only study the physical signs which accompany the states of brain which produce them.

CHAPTER IV.

MODES OF EXPRESSION BY MOVEMENTS, AND THE RESULTS OF MOVEMENT.

Movement a physical and visible action; it is often observed in physiological inquiries; it is correlatable with other modes of force—A movement expresses the action that produces it—Examples of expression by movement: anger, laughter—Results of movement—Expression by the voice, apparatus of porcupines, stamping of rabbits—Secondary movements—Work done the result of movement—Posture as a result of movement—Subsidence of movement in sleep, in fatigue, and when the attention is attracted—Spontaneous and voluntary movements—Movements of a bee from flower to flower—Summary.

COMMON experience shows that the manner and kind of movements seen in a man are expressive. Movements may be expressive of the fact that a child sees, and hears. We know that a child hears a whistle, because he moves at the sound; or if he sees a light, this may be indicated by the fact that he turns his head towards it. The movements are expressive of the subject hearing and seeing.

Movement as a physical sign, or mode of expression, appears to me a result of the properties and functions of the subject that is particularly worthy of extended and accurate study. Movement itself is a physical mode of expression, capable of very accurate observation both in time and in quantity; it is capable of being easily noted by more than one observer at the same time, and can be recorded by various adaptations of the graphic method. Such records can be preserved, and submitted to analysis by mathematical procedures. Observations of this kind have been largely employed by physiological experimenters, and much accurate knowledge has been thus obtained, elucidating processes of circulation, respiration, reflex action, etc. Movement, however produced, whether by vital action, nutrition, or otherwise, can always be shown to be correlatable with other modes of force. It is surely reasonable to observe, record, and study, movements resulting from vital action as an expression of that vital process, and not to confine observation to histological or structural (trophic) effects of vital action. It will be granted by the reader, that movement always indicates some force antecedent to that movement; it follows that movement in a body is expressive of a force acting upon, or in, that body coincident with the observation of the movement, or antecedent to it.

Movements may, then, be studied as modes of expression. In the common experiences of life, movements are accepted as expressive of conditions in the subject, as is easily shown by the analysis of examples. Let us look and inquire, then, whether movements can be shown to be a mode of expression in accordance with the meaning of that term given in chap. ii.

It is said that a manner of movement is expressive. If a movement is expressive, or if, in other words, there is expression in movement, we ought to be able to show what is the condition hidden in the organism corresponding to the visible movement. If it can be shown what produces the movement, it is shown what is directly expressed by that movement. It is shown in chap. vi., which deals mainly with questions of physiology, that movements may result from stimulation of the central nerve-mechanism, and it is shown inferentially, that movements correspond to the action of the central nerve-mechanism. It is the fact that the movements we observe are, we believe, produced by the action of the central nerve-mechanism that makes us look upon movements in the body as expressive of the inward condition. Movements are the direct expression of the action of the nerve-mechanism. The terms "manner of movement" and "kind of movement" are analyzed and explained in chap. v.

That expression may be produced by movements, and that it is most commonly so produced, can be demonstrated by examples.

Anger is commonly spoken of as an emotion or passion of the mind. Bain* says, "The physical manifestations of anger, over and above the embodiment of the antecedent pain, are (1) general excite-

* "Mental and Moral Science," 1872, p. 261.

ment; (2) an outburst of activity; (3) deranged organic functions; (4) a characteristic expression and attitude of body; and (5) in the completed act of revenge, a burst of exultation."

Sir Charles Bell * gives the following description: "In rage the features are unsteady. The eyeballs are seen largely; they roll and are inflamed. The front is alternately knit and raised in furrows by the motion of the eyebrows; the nostrils are inflated to the utmost; the lips are swelled, and, being drawn by the muscles, open the corners of the mouth. The whole visage is sometimes pale, sometimes turgid, dark, and almost livid; the words are delivered strongly through the fixed teeth; the hair is fixed on end like one distracted, and every joint should seem to curse and ban."

Henry Siddons† illustrates the expression of anger thus: "Thus, as I have been saying, *choler* adds energy to all the exterior parts of the body, but chiefly arms those most proper to seize, attack, or destroy. Swelled by the blood and humours which are thither carried in abundance, they agitate themselves with a convulsive violence. The inflamed eyes roll in their orbits, and dart forth fiery glances; the hands and teeth manifest a kind of interior tumult, by the grinding of one and the agitation of the others. It is the same kind of eagerness which the mad bull and furious bear display, to make use of the arms with which nature

* "Anatomy of Expression," p. 177.
† "Practical Illustrations of Rhetorical Gesture and Action," p. 118.

has furnished them. Moreover, the veins are swelled, especially those about the neck and temples. All the visage is inflamed, on account of the superabundance of blood carried up to it; but this redness resembles not that occasioned by desire or love; the movements are more hurried and more violent; the step is heavy, irregular, impetuous."

These descriptions of anger, as given by the philosopher, the physiologist, and the actor, agree in ascribing much of the expression of the emotion to movement and the results of movement.

Take Siddons' description. He speaks of choler as expressed by energy in the movements of the arms, grinding of the teeth, and agitation of the hands. He tells us of the effect of other movements, or arrest of movement, which he does not describe. He speaks of the limbs as swelled with blood, and the eyes as being congested; these phenomena are secondary results of spasm in the respiratory muscles. Sir Charles Bell describes movements in the face, and likewise refers to the effects of arrested respiration.

Anger in animals is in part expressed by showing the teeth. Here we have the description of certain movements, and results of movement, said to be expressive of the emotion "anger."

Sir Charles Bell,* in his sixth essay, when speaking of laughter, says, "We have seen that the muscles which operate upon the mouth are distinguishable into two classes — those which surround and control the lips, and those which

* *Op. cit.*, p. 146.

oppose them and draw the mouth widely open. The effect of a ludicrous idea is to relax the former, and to contract the latter; hence, by lateral stretching of the mouth, and a raising of the cheek to the lower eyelid, a smile is produced. The lips are, of all the features, the most susceptible of action, and the most direct indices of the feelings. If the idea be exceedingly ridiculous, it is in vain that we endeavour to restrain this relaxation, and to compress the lips. The muscles concentrating to the mouth prevail, and become more and more influenced; they retract the lips, and display the teeth. The cheeks are more powerfully drawn up, the eyelids wrinkled, and the eye almost concealed. The lachrymal gland within the orbit is compressed by the pressure on the eyeball, and the eyes suffused with tears."

Here an exceedingly ludicrous "idea" is spoken of as an antecedent and cause of certain movements of the face and other parts. Movements have been shown to be expressive of emotions; anything which indicates the movement is equally expressive of the emotions. When we look at a watch, we are quite satisfied that the spring and the train of wheels are in motion if we see the movement of the hands, because moving of the hands is necessarily a result of the movement of the wheels within, and nothing else could cause the hands to move continuously.

In many cases the results of movements are equally expressive with the movements themselves; it may be that the result of movement is more noticeable than the movement itself. In describing the expression of anger we speak more commonly of showing the teeth than of retracting the lips, but of course the teeth are shown by movements of the lips. In the expression of rage or anger, Bell describes the visage as sometimes turgid, dark, and almost livid; this congestion is a secondary result of spasmodic contraction of the respiratory muscles of the larynx, accompanying the clenching of the teeth, leading to a condition of asphyxia. Showing the teeth, and lividity of the countenance, are, then, secondary results of movements, and are direct expressions of the condition of the individual.

The voice is another important mode of expression of the emotions by the results of movement. Dr. Foster,* in speaking of special muscular mechanisms, says, "A blast of air, driven by a more or less prolonged expiratory movement, throws into vibrations two elastic membranes, *chordæ vocales*. These impart their vibrations to the column of air above them, and so give rise to the sound which we call the voice." Alterations in the tension and position of the vocal cords, and variations in the movements of the respiratory muscles, cause the changes in the voice which are expressive of the emotions. Darwin† gives the following examples of expressive sounds produced by animals through the action of apparatus that cannot be called in any sense "vocal," because not dependent on the respiratory apparatus: "Rabbits stamp loudly on the ground as a signal to their comrades; and if a man knows how to do so properly, he

* "Physiology," p. 527.   † "Expression," p. 93.

may, on a quiet evening, hear the rabbits answering him all round. These animals, as well as some others, also stamp on the ground when made angry. Porcupines rattle their quills and vibrate their tails when angered; and one behaved in this manner when a live snake was placed in its compartment. The quills on the tail are very different from those on the body: they are short, hollow, thin, like a goose-quill, with their ends transversely truncated, so that they are open; they are supported on long, thin, elastic foot-stalks. Now, when the tail is rapidly shaken, these hollow quills strike against each other and produce, as I heard in the presence of Mr. Bartlett, a peculiar continuous sound."

Secondary movements may be expressive without producing sound. Tossing of the head is often very expressive in a girl; the movement, slight in itself, is rendered more conspicuous by the secondary movements of her long curls. So the horse when neighing shakes his mane, and tosses it in the air—movements which express freshness and vigour. A man in prostrating himself on his face bends his body forward by a voluntary movement, till the centre of gravity of his body is in front of his base of support, then the body falls, not as the result of a further voluntary effort of movement, but as the consequence of the action of gravity. The first part of this movement is voluntary, and is therefore expressive of volition; the latter part of the movement of prostration is solely the result of gravity.

Work done is a necessary result of movement of

any kind, but we need not here stop to enter upon that philosophical problem which depends upon the law of the conservation of energy. Work done as the result of movement is positive proof of the movement having occurred, therefore work done is the result of the activity of the agent that produces it. It follows that work done may be just as good an expression as is movement. This is an important principle of expression, and different examples must be examined. A certain amount of labour may be performed by a man, such as raising water from a well to a high cistern by means of a pump, or a certain area of land may be dug over in the day's work; the amount of energy spent in this labour will be some kind of expression of the man's brute strength. The number of pages of writing accomplished by a literary man is some expression of his mental toil. A fine work of art, painting, or sculpture is the expression of capacity in the artist, and it is also an indication of the quantity of his exertion in a given time.

One effect of movement in man and animals is to produce locomotion; locomotion is work done by movement. If all movements may be expressive, locomotion may be expressive of the quantity, time, and kind of work done. Is not the manner of a man's walk often highly expressive? In a man the manner of walking is characteristic of the individual. We may also find types of walk. There is the man whose rapid strides indicate his excitement, and the slow and dawdling walk indicative of purposeless aim. The step may be heavy, irregular, impetuous, or hesitating; it may be brisk, free, unrestrained, easy, and mobile.

A posture of the body is the result of movement. The term "posture" indicates the relative position of the several members of the body with regard to one another, or the relative position of the individual parts of a member. When I began to make a definite clinical study of expression, in the sense in which the subject is treated in this work, I frequently looked at my patients after diagnosis had been made, to observe if there were any outward expressions of their organic or nerve-condition. My attention was soon attracted to the frequence of certain postures of the body indicative of conditions of the nerve-system. In these early studies it was found more easy to give an accurate description of a posture than of a movement, because a posture is a condition of quiescence, a movement is of temporary duration. A posture is described when we have given an anatomical description of it, and the matter is more simple to deal with than a movement. In works of art, both painting and sculpture, it is very largely by the posture of the body that we judge of the condition of the man or animal represented. These matters are fully considered in chaps. viii., ix., and x. Now, a posture, while it is maintained, implies absence of movement in the part. When we see a man's fist clenched in passion, we may study the posture of that hand, and, as long as it is maintained in that posture without alteration, there is no movement of the parts of the hand. A posture is the result of the last movement, and it remains till the next movement occurs. Postures are expressive as the results of movement, and absence of movement is essential to the absolute equilibrium of the posture.

We now have to look at examples where absence of movement is a mode of expression, in contrast with others where movement is present. Expression may be produced by the absence of movement, as well as by movement observed. A child, after the day's work and play, is put to bed: we observe a total subsidence of voluntary movements, the eyes are closed, the muscles of the limbs relaxed, no reflex actions occur from moderate light or sound, but the respiratory movements continue with regularity; then we say the child sleeps. The subsidence of movement, other than the respiratory movement, is the principal indication of the condition sleep. A man, wearied by the bodily or mental toil of the day's occupation, sits in a chair, he moves and speaks but little, his attention is but slightly attracted by objects around; it is with difficulty that he is stimulated to movement by the voices of his children; we say that such a condition in the man expresses his fatigue and exhaustion. After rest and refreshment his expression is different, he moves with briskness, he talks to those around, he is quick to observe all about him, he plays with his children and joins in their games: here the man's movements, and the results of his movement, indicate his freshened activity.

The absence of ordinary movement may be due to other causes than exhaustion. Watch a hearty child, say two years of age, sitting on the floor with playthings around him: his movements are incessant, his head, face, eyes, hands, fingers, etc., are incessantly moving. If a stranger enters the room the child stops his play, his various movements cease, his eyes are cast down, scarcely any indication of his former condition of movement remains. There is no reason to suppose that the general conditions of his nutrition are interfered with, for the circulation and respiration continue as heretofore. His various movements are arrested or inhibited at the sight of the stranger, and this inhibition of movement is characteristic or expressive of the child's condition. When any object strongly attracts the attention of a man it usually produces inhibition of movements. This is probably a very important factor in some mental phenomena.

Nutrition may be directly expressed by movements, for it commonly happens that movement is the most obvious outcome of nutrition in a subject. A young infant is full of movement while awake, if its nutrition is good; its arms and fingers are moved apparently spontaneously.* A baby thus lively is well nourished. Clinical experience shows that the condition of its nutrition is in part indicated by this spontaneous movement. Now, if the child becomes ill, say from the effects of bad feeding, from lung or stomach disease, the spontaneous movement described almost ceases, the lowered condition of nutrition lessens the amount of spontaneous movement. A large amount of

* See chap. xiv. fig. 31, p. 245.

spontaneous movement is equally indicative of good nutrition in animals. Note the frisky play of puppies and kittens. The sickly dog, whose nutrition is greatly impaired, can hardly lift a paw or wag his tail. In some pathological conditions movement may be in excess, and not indicative of a good condition of nutrition; but it will not be convenient to enter upon the discussion of this matter at present.*

In speaking of the modes of expression by movement, we can hardly be permitted to pass silently over the consideration of the difference between spontaneous and voluntary movements, but it must be remembered that we are not here directly concerned with the essential properties expressed by the objective signs. My reason for passing over so important a point is that a discussion as to the criteria distinguishing spontaneous from voluntary movements would involve us in a philosophical argument out of place in this work. Having carefully considered this point, I have come to the conclusion that, in place of attempting to distinguish between spontaneous and voluntary movements, it will be easier and more satisfactory to study inborn, and acquired sources of movement. Movements seen to occur upon birth and soon after, may be assumed to depend upon inborn conditions, causes inherent in the subject at birth, so that such movements may be called inborn movements. As previously said, the child upon birth presents very constant movements, which may be conveniently spoken of as inborn in contradistinction to the

* See chap. vii. p. 125.

acquired and voluntary movements of adult age. The great distinction between the inborn and acquired movements in kind, is that the acquired movements are found in each successive year of the child's life to be more and more co-ordinated; they become more easily co-ordinated into complex arrangements by the temporary action of some external force.

I have, at present, but little exact and scientific knowledge as to the differences in kind between the inborn and the acquired movements, but this does not prevent us from considering the question; and the consideration and discussion may be an encouragement to the investigation of the problem by scientific methods. The expression of inborn properties, as distinguished from acquired properties, is a very important and interesting subject. In an investigation for the purpose of determining what movements are inborn and what acquired, it would be necessary to observe the spontaneous movements of young children,* also to observe what reflex actions can be excited in them, and compare these movements with analogous conditions in later life. These questions, were we conversant with them, would supply some information as to the modes and processes of evolution in the individual and in a species; we should also gain further knowledge as to expression by co-ordinated movements, a kind so very characteristic of mind. This subject will be further developed and defined in chap. v. and further explained and illustrated in chap. vi.

* See chap. xiv.

Expression of the relations between two or more subjects may be effected by the movements of some kind of intermediate agent, some kind of movement between the subjects that is an expression of the condition of either. If we look at a bank of flowers on a fine summer day, we observe the bees flying from flower to flower. Usually a bee visits only one species of flower in the same journey, and his movements indicate something about himself and about the flowers—a relation between bees and flowers that has been well put forward in the writings of Springel, Kerner, and Darwin. Looking at an ants' nest a wonderful order may be observed in the movements of the little creatures, each appearing to perform its own part in an organized system which appears to be arranged for the benefit of the whole community and not of the individual. It is the movements of the ants and the work accomplished by them which indicates the organization of the whole, and the so-called instinctive properties of each. In life in a city we see men pass from house to house, from home to office, etc., these movements, as observed, proving social organization dependent upon the make of each man and the relation of men to one another. When we learn of the movements of an army in the field, and hear how the different portions separated from each other, work in unison for the accomplishment of one end, we find that the relation of the movements of the parts form an expression of the government of the whole. Now, this mode of expression by movements is one of the highest which we have to consider, involving, as it does, many subjects, and many forces. Still, the detailed consideration of each of the examples given shows the possibility of understanding what is expressed by the movements in each case.

A bee visiting flowers expresses certain relations between these subjects. The bee is acknowledged to be impressionable, to be guided to the flower by the sight of the flower, *i.e.* by the light reflected from the flower. The light reflected from the flower, acting upon the impressionable bee, directs its flight and movements. Is the flower impressionable? Is the flower affected or impressed by the insect at its visit? It is certain that variations in plants, and in their flowers in particular, do occur, fitting them to the visits of particular insects. Are these adaptations solely due to spontaneous variations preserved as being the fittest? Is not the modification of the flower helped on by each visit of the insect, to which it ultimately becomes so well adapted?

In conclusion, let me quote from a speech by Sir William Gull[*]: "A tone of the voice, the play of the features, the outline and carriage of the body, are to him (*i.e.* the physician) as invariably related to the central conditions which they reveal, as are the grosser facts of Nature."

Summary.—Movements in a child may express that it can see and hear. Movement as a mode of expression is particularly worthy of study. It is capable of measurement in time and quantity, and can be recorded by the graphic method; the records can

[*] Inaugural address, International Medical Congress, 1881.

be analyzed; further, all movement can be correlated with other modes of force. For these reasons it is urged that kinetic function, as well as trophic function, should be studied in all living beings. Any movement tells us something about the source of the motor force, which in man is the central nerve-system. The results of brain-conditions, called the emotions, are expressed by movements; this may be illustrated by quotations from the writings of Sir C. Bell, Bain, and Siddons, in their descriptions of laughter and anger.

Anything that indicates movement may be as expressive as the movement itself. Movement in the hands of a watch indicates the movement of its wheels. We say that in anger a man shows his teeth; we really mean that he opens his mouth and moves his lips. In the expression of rage, the congestion of the face results indirectly from fixation of the respiratory muscles. So the voice results from movements, and is a mode of expression. Similarly, work done is a result of movement, and may be a mode of expression, whether the work be mechanical or mental in kind. The gait in walking indicates a man's general condition, eagerness, excitement, fatigue, etc.; it depends upon the condition of his nerve-centres, and expresses this.

A posture is the result of the last movement of the part. The significance of postures is shown in chap. vii.

The subsidence of movement may be as suggestive a mode of expression as its occurrence. Spontaneous movement is lost in fatigue, and in exhaustion. The spontaneous movement of infancy subsides in sleep, and on the occurrence of organic disturbance.

In certain cases, the movements of two or more independent subjects, not connected by any material link, are expressive, as the movements of a bee from flower to flower, or the movements of men in social life.

CHAPTER V.

MOVEMENTS AND THE RESULTS OF MOVEMENTS CONSIDERED IN THE ABSTRACT, OR APART FROM WHAT THEY EXPRESS.

Movements are means of expression—Movements classified as reflex, voluntary, spontaneous—The attributes of a movement are its quantity, kind, and time—Time of a movement most conveniently recorded by the graphic method—Frequency and duration, the importance of considerations as to time—Two movements, considered in relation to time, may be synchronous; this may depend upon an organic union of the motors, or upon each working in similar rhythm—Expression may consist of coincidences or combinations of movements—The number of possible combinations of synchronous movements of n subjects is $2n$; the number of sequences of such combinations is unlimited—Actions described as a series of movements—Description of a dog in terms of movement and growth—Co-ordinated and inco-ordinated movements—Walking described as a series of movements—Movements of an aggregation of independent individuals—Principles of analysis of movements—Description in anatomical terms—Contrast of movements of small parts and large parts of the body in their physiological significance—Interdifferentiation—Collateral differentiation of parts—Symmetry of movements, indicating like action on both sides of the brain—Asymmetry of movements common in the higher functions—Classification of movements: according to anatomical analysis; according to the physiological principles of analysis given above; as intelligent and non-intelligent; as synchronous or non-synchronous; as occurring in regular series; as accompanied by feelings, other classifications are suggested—Summary.

MOVEMENT as a metaphysical abstraction we need

not consider; movement as a property of the subject considered is what we are here concerned with. In this chapter we commence with the admission that movements may be expressive, an admission which leads us to study movements apart from their cause, for the sake of simplification of the subject. Movements of various kinds are the principal modes of expression, as is illustrated in the foregoing chapter. Indeed, it is probable that it may be shown hereafter that expression by coincident development * is only a mode of expression by the results of molecular motion. Movements may be classified according to their mode of production, or our ideas thereupon; thus, in a man, certain movements are said to be spontaneous, others reflex, others voluntary.

Again, the classification may be made according to the attributes of the movements—kind, quantity, and time; which latter attribute includes frequency, speed, duration. Other movements may be continuous or interrupted, and they may be considered in one or in more subjects.

As to the movements of a single subject. In all physiological work any calculation concerning movement is best made by recording such movement on paper by some graphic method. Now, in looking at a tracing thus produced, we see certain indications in the form of the outline of the curves; if the recording paper always travels at one uniform rate, the form of the outline will be plainly characteristic of the kind of movement observed in the same subject

* See chap. xvi.

on different occasions. Thus the regular movements of the heart give a regular tracing on paper; tracings obtained by movements of the fingers in chorea are most irregular. By "the frequency of movements" we mean the number of observable movements in a minute or other division of time. The term "duration of movements" is self-explanatory. "Rate" or "speed" implies the degree or length of movement accomplished in a given time. The term "quantity of movement" is used in the sense of quantity as correlatable with other modes

Fig. 2.

of force, as used in speaking of the law of conservation of energy. These considerations regarding movement show at once how important time is as an attribute of a movement.

Now, if two or more movements be considered in relation to time, further points arise expressive of the relation between the times of the one movement and of the other. Two movements may be synchronous throughout, as when two trains run on lines side by side, and start at the same time, and continue to move at the same rate. If two men run in concentric circles at the same rate the movement of their bodies is synchronous, and they maintain their same relative positions; if their rates are unlike, one gets an advance of the other, but after a while they meet again, and the times of the successive meetings or coincidences is a calculable matter if the rates of movement are known.

Turning back from these abstract considerations to our proper subject, expression by movements, a few examples will show how the one bears upon the other. In discussing one mode of expression of the hands, we say, " The fingers of the hand are opened and closed in rapid sucession, indicating the excited and angry feeling of the man." Now, here the movements of five subjects, five digits, are said to coincide, they move synchronously, opening and closing. Now, such synchronous movements may result from the equal pace at which each finger tends to move, or they may result from an organic connection between the cause of each movement. The same facts of movement may be differently expressive, according to the causation of the synchronous action. Problems may arise, then, in considering modes of expression, in which the expression is described in terms of coincidences or combinations of movement. A series of movements may occur in any one subject, or, considering two or more movements as regards their time, a series of combinations may occur in definite sequence. The number of possible combinations of synchronous movements is strictly limited by the number of movements considered; if the number of movements considered be n, the number of possible synchronous combinations of such movements is represented by $2n$; the number of variations in the order of the sequences is without limit, or infinite.

As an illustration, consider the movements of the five fingers. Let the five digits be respectively represented by the letters A, B, C, D, E, and let us consider the combinations in which they move together. There are thirty-two possible combinations of such movements, viz., A, B, C, D, E, AB, AC, AD, AE, BC, BD, BE, CD, CE, DE, ABC, ABD, ABE, ACD, ACE, ADE, BCD, BCE, BDE, CDE, ABCD, ACDE, ABCE, ABDE, BCDE, ABCDE, 0; but the number of possible variations in the order of sequence of combinations is infinite. These facts illustrate how the number of postures of the hand, *i.e.* coincident positions of its parts, is finite, but the variations of expression by variation of its movements are infinite.

Many acts or actions may be described by recording the movements, combinations of movements, and sequences of movements. A clear understanding on this point is essential to many of the arguments to be brought forward in this work, especially to understanding the importance attached here to the accurate and detailed study of movements, and the full appreciation of what such studies may teach us as to methods of investigation, and the methods of evolution. Analogies will be made hereafter between combinations and sequences of trophic action, and of movement.

In order to make our ideas more clear we will take one illustration in some detail.

The puppy of a setter-dog at its birth is a very different animal from what it is at an adult age. The difference between the two stages of growth would at once strike any ordinary observer, and, if he analyze and arrange the points of difference, he will find differences in size, form, proportions of the body; also differences as marked in the behaviour, running, steadiness, and kind of work done, *i.e.* differences in the kind of movements done and the results of movement. Thus analyzing the points of difference between the puppy and the full-grown dog, it will be seen that it is as important to describe the movements as an expression of the difference between youth and maturity, as to describe the changes in the body alone. Further, there are other changes in function in the animal as the puppy grows to be an old dog; the conditions of its movements change, and its body, its *corpus*, changes independently of its movements. There is seen, then, an alteration in two kinds of functions, or properties, as the animal grows old: first, a change in the functions of movement; second, a change in its trophic condition. A complete historical account of the growth or change of a puppy into an adult animal requires the description of the changes in the conditions of its movements and the differences in its trophic conditions, or growth.

It is very usual to speak of the important differences between co-ordinated movements, and inco-ordinated movements. In this chapter we have nothing to do with the causes of co-ordination of movements in man or in animals, but we have to establish rules by which we may distinguish the co-ordinated from the inco-ordinated movements. When a man walks steadily along a road we see that his movements are regular, there is a similar repetition of events, and similar impressions are left at equal intervals upon the dust of the road; such a man's movements in walking are said to be co-ordinate, and the regularity in the successions and combinations of his movements, as indicated by the general regularity of the movements of his whole body and the uniformity of his foot-prints, indicates to us the perfect co-ordination of his movements in walking.

Contrast the movements of such a good walker with those of a man afflicted with the condition termed locomotor ataxy, and contrast the foot-prints left by the two men. In locomotor ataxy the patient walks with a precipitate gait, and staggers, the legs start hither and thither vaguely, and the heels come down at each step as in stamping. Such movements are said to be inco-ordinate. Our present object is to inquire how we can best define and formulate what we mean by co-ordination. This, I believe, can only be done with scientific exactness by definitions framed in terms of combinations and successions of movements and results of movements.

Locomotion, or walking, has been described and experimented upon in man by M. Marey.[*] He says, "The most simple and usual pace is *walking*,

[*] "Animal Mechanism," p 111.

which, according to the received definition, consists in that mode of locomotion in which the body never quits the ground. In running and leaping, on the contrary, we shall see the body is entirely raised above the ground, and remains suspended during a certain time. In walking, the weight of the body passes alternately from one leg to another, and as each of these limbs places itself in turn before the other, the body is thus continually carried forward.

"This action appears very simple at first sight, but its complexity is soon observed when we seek to ascertain what are the movements which occur in producing this motion."

This shows us that we may define walking as a succession of paces. "A pace is the series of movements executed between two similar positions of the same foot." Now, a series of movements implies something tolerably simple. Marey tells us, then, that walking is a succession of paces, that a pace is a series of movements. It remains, then, for us to inquire what properties may be presented by a series of movements.

Walking may be described in various ways. As proposed already, it may be described as a series of movements of the parts (limbs) which produce the locomotion; as a series of muscular contractions moving the limbs; or, did we possess sufficient knowledge, we might describe a pace in walking by stating what nerve-centres are in action coincidently, and what is the order of their successions.

M. Marey has in part described locomotion by

indicating the movements of the trunk in walking as resulting from the movements of the limbs, *i.e.* he describes the movements of the trunk as moved by the limbs, as moved by the muscles, as stimulated by the nerve-centres. If movements occur in a subject it is implied that the parts of that subject can move more or less independently of one another. When we say a man's limbs move, it is implied that his limbs are movable. If a billiard-ball upon a table is struck and moves in consequence, the ball as a whole travels, the movement that follows is that of the whole ball. In speaking of movement it is, then, necessary to be clear whether we are speaking of movement of the object as a whole, as when we speak of the movements of the moon, or whether we mean movements of the parts of the object, as movement of the limbs of an animal. It often happens that a certain movement that is observed, is the movement of an aggregation of independent individuals. Such is seen in a procession of men assembled for a political purpose. The movement of the whole procession is an aggregation of the actions of a number of separate men, held together and governed for the time by a common political object; there is no structural union between them: and in such a case the coincidence of their movements is an expression of the bond of union. The movements of such an aggregation are often highly expressive, the more so from the fact that there is no organic union between the individuals. The movements of an army corps are more expressive than the movements of a gang of slaves chained together—the cause must be stronger which governs the body of free men. The movements of an army may be expressive of the condition of army discipline, the will of the general, the orders of the Home Government, acting under the influence of Parliament and the nation. In any example of expression by movement it is, then, necessary to distinguish action of an aggregation of separate individuals from a collection of organically united individuals.

Man is the special subject with whom we are concerned, and I wish now to put forward a brief scheme for the *analysis of movements* seen in man; this scheme is founded upon experience in medical practice.

When we see movements in a man, we may proceed to define them[*] in anatomical terms. Thus, if a man holds out his arm and hand straight and level with his shoulder, any movement in the parts is easily observed. A movement may be seen to occur, and may be indicated by the following terms: Flexion of wrist, flexion of all the metacarpo-phalangeal joints, strong flexion of the thumb at each joint, the thumb being thus brought into the palm of the hand with the fingers bent over it; phalanges adducted, metacarpal bones adducted, thus arching the palm of the hand.

Here we give an *anatomical description* of a movement that might be observed. It is necessary in making an observation, to record the movement, or series or combination of movements, thus, before

[*] See tables of analysis, chap. ix.

any attempt can be made at analysis or classification; and we must keep a record of the movement before us while we seek explanation of the phenomena and try to understand what is expressed by the movement.

After making an anatomical description of the movement for the sake of keeping it before us as a mental account of the transient phenomenon, we may apply certain *physiological principles* to the analysis of the record of the movement.

It is well known to physicians that there is much practical difference in the signification of *movements in larger parts of the body of man in contrast with movements seen in the smaller parts.** It is necessary in a compound movement—a general movement compounded of the movements of different parts or subjects—to distinguish in description the action of small parts, such as the muscles of the eyes, face, fingers, from movements of larger parts, such as the skull, the shoulder, elbow or wrist, the hip or the knee. This principle of the different physiological signification of the action of small and large parts will be illustrated in chaps. vii. and viii.; so also with the other physiological principles.

I think there is a considerable importance in the two next principles, though I believe they have been less generally studied than they ought to be. These principles I name "Interdifferentiation" and "Collateral Differentiation."

The term *"interdifferentiation of movements"* is

* See chap. vii. "Small parts" are most affected in hemiplegia.

used to indicate that the conditions of movements are different in the large and in the small parts. There may be a great difference observed, in the upper extremity of a man's arm, in the amount of movement, or in the number of movements of the small parts as compared with the large parts. Thus, in writing, painting, sewing, the fingers, knuckles, and wrist do almost all the movements, there being comparatively little action in shoulder and elbow: here, then, is marked interdifferentiation of the movements of the upper extremity. Contrast this example with the use of a man's upper extremity for ground-digging or boxing. In this case the fingers and small parts are but little used, the movement is from shoulder and elbow; here, again, interdifferentiation of movements is marked in degree, the large parts being more used than the small parts.

Collateral differentiation of movements, or a difference in the movements of collateral parts, is our next principle. The knuckles and the fingers are collateral parts: they are all equally small parts, they can move all together or each can move separately. There may be differences in the movements of the fingers, or collateral differentiation of movements may be seen. In writing, the movements of the pen are effected almost entirely by the thumb, index, and middle fingers: this is collateral differentiation of the movements of the fingers in the act of writing.

Symmetry of movement in man and in animals is by no means a new subject; it has been noted and often discussed. I think that symmetry of movements, in a physiological sense, indicates equal bilateral movements, movements occurring similarly on either side of the body. The physiological importance of symmetry is very different in various parts of the body: the expression indicated by a symmetry of hands is very different from the signification of a symmetry of the face. Here we may stop one minute, again to point out an analogy between conditions of movement and conditions of trophic action. Symmetry of movement and symmetry of growth and structural development are seen in man in the nutritive organs, lungs, and respiratory movements; the two halves of the brain in their anatomical structure and the two sets of nerves with organic functions, vagi, phrenics, sympathetics, are fairly symmetrical.

Asymmetry in man chiefly concerns his higher functions; it is seen in his movements in writing, and other high-class hand-work. Speech is not equally represented in either hemisphere of the brain. Some passions are expressed asymmetrically in the face. Evolution is often expressed in plants by asymmetry of growth (see chap. ii.).

Classification follows naturally after analysis. I have purposely omitted to consider all the means of analysis of movements, as such might prove wearisome, and is hardly necessary to clear views on the general subject of this work, "The Modes of Expression." For the same reasons our notice of classification will be but brief, but both analysis and classification will be abundantly illustrated in succeeding chapters. In any special problems as to a mode of expression and its meaning, it is of course necessary to use all methods of analysis, and also to see in what class the special mode of expression under consideration is to be placed.

The following classifications of expressive movements are suggested as practically useful :—

1. Anatomical analysis gives a method of classification; the same anatomical movements will always appear in the same class under this arrangement. It must be understood, however, that similar movements have a very different signification under different circumstances. Closing the fingers when grasping an object, when in passion, and in an epileptic fit, may be a similar movement, but the signification varies in each case.

2. Movements may be classified according to the physiological principles given.

3. Movements are often spoken of as intelligent and non-intelligent. This division is different from those used above, and does not depend solely upon their analysis; their signification depends upon the previously ascertained truth, or uniformity, that certain kinds of movement are only produced by a central nerve-mechanism which can also produce "intelligence" or non-intelligence at the same time as it produces the particular movements in question. Good-class painting, high-class manipulations are expressive of intelligence, because they are uniformly found to accompany intelligence. Walking is not necessarily accompanied by intelligence: a man may walk in his sleep, an idiot may walk.

Again, sucking in an infant is not a sign of present intelligence. Some movements are usually accompanied by non-intelligence.

4. Movements may be synchronous or asynchronous.

5. Movements may be classified as occurring in a regular series. The regular movements of a particular dance may be expressive. A series of movements is often expressive of certain passions, emotions, or states of feeling. In some kinds of fits a series of movements characterizes the attack.

6. Movements are often spoken of as caused by feelings. It seems to me that this is a great assumption. It may be, on the contrary, that the movement and the feeling are alike due to the condition of nerve-centres.

Summary.—For the sake of simplicity movement is here considered apart from the cause of that movement which it directly expresses. Various modes of classification of movements may be used— as spontaneous, reflex, voluntary; this classification depends upon our idea of their origin. A classification may be made according to the attributes of the movement—frequency, speed, duration, quantity, etc. The graphic method is most suitable for such a classification. Further, description of the attributes of a movement involves consideration of the analysis of tracings of movements; the consideration of the attributes of a movement involves the laws of conservation of energy, also ideas of the relation of time, and quantity. When the relations of two or more movements are considered, intricate mathematical problems arise which are probably more complicated than questions in astronomy. The possible number of combinations of movements is finite, as is easily illustrated. It appears probable that a very direct analogy may be made between series of movements, and series of acts of growth, and that similar laws may govern both.

CHAPTER VI.

PHYSIOLOGY OF EXPRESSION.

Modes of movement in plants; in the amœba; the ascidian has a nerve-mechanism, and apparatus for reflex movements—Nerve-mechanism of vertebrates—Nerve-muscular apparatus; nerve-muscular action—Do certain nerve-centres produce certain movements?—Ferrier's experiments—Cerebral localization—Nerve-centres—Visual perception indicated by movements—Time requisite for a reflex movement—Inhibition of movement—Physiological effects of light, in man, in plants—Light stimulates trophic and kinetic action—Effects of light in the new-born infant; movements stimulated, inhibited, co-ordinated—Retentiveness to effects of light—The brain of an idiot not thus impressionable to light—Summary of the effects of light—Extrinsic stimuli, mediate and immediate—Trophic action of light—Summary of effects of light on plants.

In dealing with the physiology of expression we are concerned solely with facts in the history of living beings, plants and animals, including man; the sections of physiology that we are mainly concerned with are those involving questions of movement and nutrition. It is convenient, then, to our purpose that we should take a brief review of the modes in which movement is produced in the organism of the lower, and higher, living beings.

Something has been said in chap. ii. about the

causes, and the mechanism of movement in plants. In animals we see movement result from the action of the organism as a whole, or from the action of parts of it, and this is one of the most constant facts seen in the lowest and highest grades of animals. The movements of amœbæ and other protozoa have been often observed. The body of the amœba is less firm than jelly, yet it has the power of moving from place to place. At first appearing as a shapeless mass, it may rapidly throw out filaments, or projections, which may again be retracted into the mass of the amœba. All parts of this mass appear to be alike, and similarly endued with contractile powers. These filmy patches glide about although the eye can detect no differentiation of their parts, and there is no special mechanism for producing force, apart from the portions that produce movements.

In such animals there is no differentiation into force-producing parts, and motor-organs; every part of the body can perform all the functions of the whole organism: thus it is seen that the amœba has an organization lower than that of the *Oxalis* and *Mimosa pudica*, which have special motor-organs.

In the *ascidian* there is a mechanism for the production of motor-power, and a contractile apparatus as a moving instrument. "The nervous apparatus consists of a central ganglion, Fig. 3 (*c*), connected with the periphery by two sets of nerve-filaments. One set is distributed to a part of the integumentary surface capable of receiving and being acted on by external stimuli (*a*), the other is distributed to muscular fibres, which on contraction cause diminution of the body cavity (*d*). Impressions made on the sensory surface are conveyed by the afferent fibres to the central ganglion,

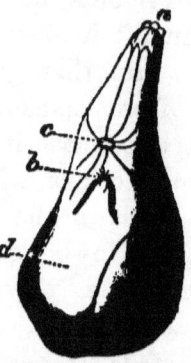

Fig. 3.—Nervous system of an Ascidian (Carpenter). *a*, the mouth; *b*, the vent; *c*, the ganglion; *d*, the muscular sac.

whence an impulse is sent along the efferent fibres, causing contraction of the muscles. Such an action is termed a *reflex action*, a term derived from the reflection, as it were, of the afferent impression back to the periphery." * The ganglion produces the motor-power; the muscular fibres are the motor-apparatus, acting when they receive an impulse from the ganglion.

Now, passing up to the vertebrates. Here we find a well-defined portion of the body, the nerve-mechanism which, when properly nourished, produces motor-force; there are also well-defined muscles, connected by nerve-fibres with the nerve-centres. When a stimulus passed from the nerve-

* Ferrier, *op. cit.*, p. 16.

centres to the muscles they contract, and produce movements. Here, then, in the higher animals, the nerve-mechanism and the muscles produce the movements, and the joint action of these two portions of the body is necessary to the production of movements; this has been proven by numerous experiments in many subjects.

The kind of movements that we have to do with in the study of expression in man are nerve-muscular signs, and it is essential to obtain clear ideas on this point, in order that the principles and modes of expression in man may be understood. The simplest nerve-muscular apparatus consists of a muscle, a nerve-centre or collection of nerve-cells, and a nerve-fibre conveying currents from the nerve-centre to the muscle; the whole must, of course, be duly supplied with blood and nourished. Force is generated in the nerve-centre by its nutrition, and is conveyed to the muscle by the nerve-fibre. When the centre sends force to the muscle it contracts; the muscle serves as an index showing the times of discharge of force by the centre. It is not necessary here to give any detailed account of the structure and arrangement of the nerve-centres; suffice it to say they are situated in the brain and spinal cord.

It is, then, convenient for our purposes in the study of the principles and modes of expression to speak of "nerve-muscular action" or signs. In the study of the brain and spinal cord we know when a current of nerve-force is sent out from a nerve-centre by the effect of that current on muscular fibres. The muscle is the index of the nerve-current; the muscular contraction is the expression of the motor action of the brain, indicating the time, frequency, duration of the nerve-current produced by the central organ.

If the central nerve-system is the organ generating the force, and stimulating the muscles to

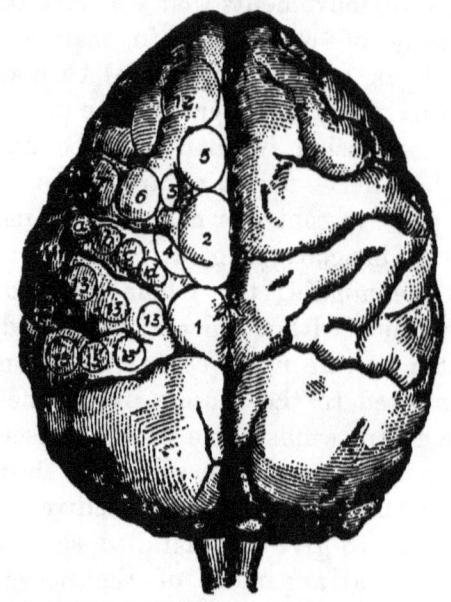

Fig. 4.—Upper Surface of the Hemispheres of the Monkey. The circles and included numerals are explained in the text (Ferrier).

contract, can different portions of that mass act separately? Do certain parts of the nerve-mechanism stimulate special movements? Are certain movements indications of the activity of certain parts of the central nerve-mechanism?

Dr. Ferrier experimented upon the brains of dogs and monkeys with the following results:—

Figs. 4 and 5 represent the lateral and upper surfaces of the brain of the monkey; the circles and included numerals indicate certain areas on the convolutions of the brain. Figs. 6 and 7 represent similar views of the human brain, and the circles with their numerals indicate areas corresponding approximately with the areas on the monkey's brain similarly indicated. In his experiments Ferrier examined each area in two ways. The portion of brain experimented upon was excited

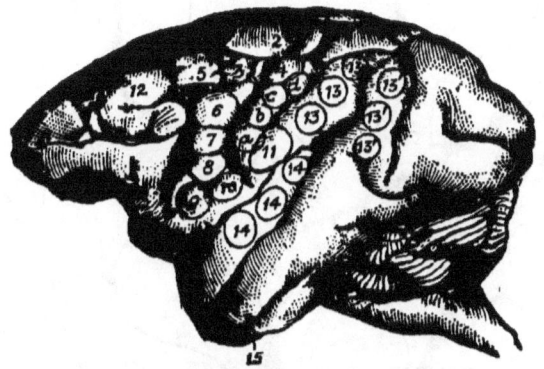

Fig. 5.—The Left Hemisphere of the Monkey. The circles and numerals have the same significance as in last figure.

by an electrical stimulus, and the muscular movement in the body which resulted was carefully observed and recorded; as a converse experiment, the same brain area was then destroyed, and the set of movements in the body which were previously stimulated were now found to be paralyzed. This demonstrated that stimulation of a certain brain area produced a certain set of movements in the body, and that when the same brain area was destroyed, that set of movements was paralyzed.

88   PHYSICAL EXPRESSION.

As the result of such methods of inquiry, the following list * was drawn up, indicating the brain areas corresponding to certain movements :—

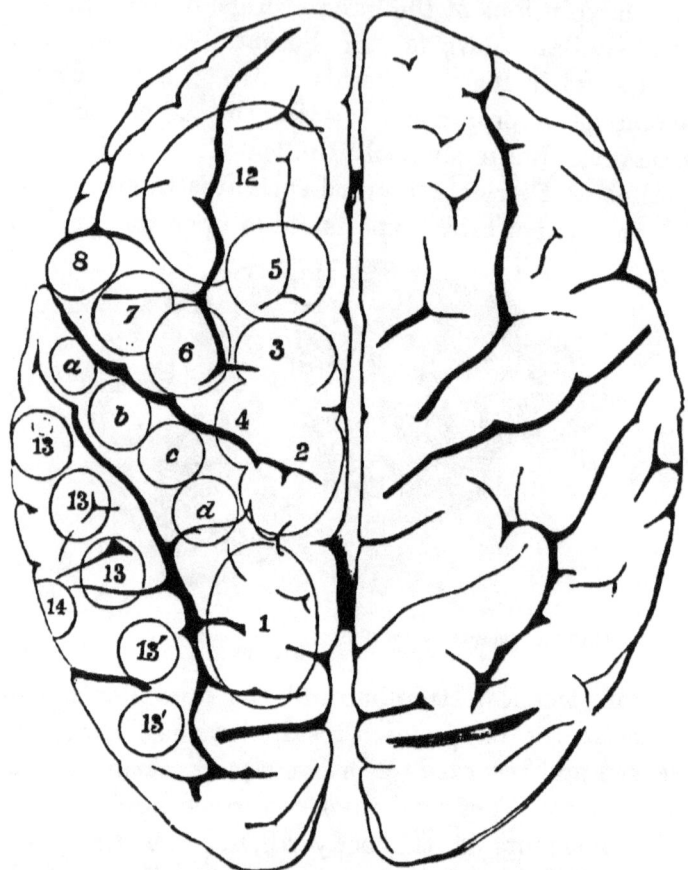

Fig. 6.—Upper Surface of the Human Brain. The circles and letters have the same signification as those on the brain of the monkey (see Fig. 4).

"(1), placed on the postero-parietal lobule, indicates the position of the centres for movements

* The brain-centres referred to in the succeeding paragraphs are indicated in the figures by the corresponding numerals.

of the opposite leg and foot, such as are concerned in locomotion.

"(2), (3), and (4), placed together on the convolutions bounding the upper extremity of the fissure

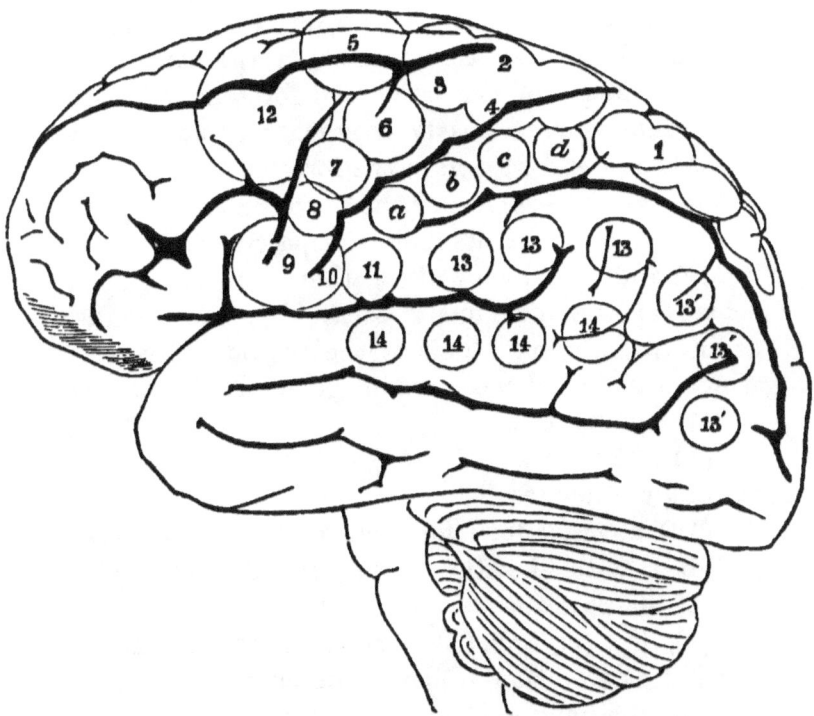

Fig. 7.—Lateral view of the Human Brain. The circles and letters have the same signification as those in the brain of the monkey (see Fig. 5).

of Rolando, include centres for various complex movements of the arms and legs, such as are concerned in climbing, swimming, etc.

"(5), situated at the posterior extremity of the

superior frontal convolution, at its junction with the ascending frontal, is the centre for the extension forwards of the arm and hand, as in putting forth the hand to touch something in front.

"(6), situated on the ascending frontal, just behind the upper end of the posterior extremity of the middle frontal convolution, is the centre for the movements of the hand and forearm in which the biceps is particularly engaged, viz. supination of the hand and flexion of the forearm.

"(7) and (8), centres for the elevators and depressors of the angle of the mouth respectively.

"(9) and (10), included together in one, mark the centre for the movements of the lips and tongue, as in articulation. This is the region, disease of which causes aphasia, and is generally known as Broca's convolution.

"(11), the centre of the platysma, retraction of the angle of mouth.

"(12), a centre for lateral movements of the head and eyes, with elevation of the eyelids and dilatation of the pupil.

"(a), (b), (c), and (d), placed on the ascending parietal convolution, indicate the centres of movement of the hand and wrist.

"Circles (13) and (13'), placed on the supra-marginal lobule and angular gyrus, indicate the centre of vision.

"Circles (14), placed on the superior temperosphenoidal convolution, indicate the situation of the centre of hearing."

When we consider that any expression by movement, or the result of movement, in man or in an animal, is an indication of activity in the central nerve-system, and is probably due to the activity of a certain part only, we find it convenient to recognize this in our nomenclature. The term *nerve-centre* is here used to imply a portion of the nerve-mechanism, which can act more or less independently of the rest, and which by its activity produces certain, more or less definite motor or other effects, observable and separable from those of the rest of the central mass of the nerve-mechanism.

This paragraph implies the belief that there are nerve-centres, as above defined, existent in the brain. Ferrier's experiments and numerous other facts tend to support that view.

In speaking of nerve-centres in the brain, we do not here wish to express the belief that these necessarily exist as localized anatomical areas of brain tissue separate from the surrounding nerve-mechanism, although this may be true. I think that we have framed a definition without implying such anatomical isolation as an essential fact.

An hypothesis is useful, if it leads to systematic experimentation and the record of observations, and I trust to be able to show that certain propositions may be defined and investigated by means of this working hypothesis that nerve-centres do exist as defined above.

In Ferrier's * experiments the sensory centres in the brain were sought for by observing that movements, or absence of movement, indicating blindness

* *Op. cit.*, p. 165.

followed in an animal when a portion of the brain called the angular gyrus was destroyed.

"In support of these conclusions, the following details selected from the protocols of the several experiments recorded elsewhere, will be sufficient. In the first experiment the angular gyrus of the left hemisphere was destroyed, the left eye was securely bandaged, and the animal allowed to recover from the state of chloroform narcosis. After recovery it began to *grope about* a little *in loco perfectly alert*, but would not *move* from its position. It did not *flinch* when held close to the gaslight. Placed in a cage beside its companions, it took no notice of them, but *sat still*. Hearing and other senses remained unaffected, and stimuli of these senses caused *active reaction*."

The quotation is given exactly; the italics are mine, and it is at once seen that the evidence of blindness was the condition of movements, the kind of reflex action following or not following upon stimulation. This affords further evidence as to the truth of the statement that we are dependent upon the motor functions of the nerve-centres for almost all the knowledge we have as to their functions.

Reflex Action.—The apparatus necessary for a reflex action has been briefly described in chap. iii.; something more must now be said concerning the time consumed in the production of a reflex movement. Foster gives the following experiment (*op. cit.*, p. 471): "If, in a brainless frog, the area of skin supplied by one of the dorsal cutaneous nerves be separated by section from the rest of the skin of the back, the nerve being left attached to the piece of skin and carefully protected from injury, it will be found that slight stimuli applied to the surface of the piece of skin easily evoke reflex actions, whereas the trunk of the nerve may be stimulated with even strong currents without producing anything more than irregular movements.

"In every reflex action, in fact, the central mechanism may be looked upon as being thrown into activity through a summation of the afferent impulses reaching it."

The change in the nerve-centre which follows upon the afferent stimulus requires a certain time, and this is proven as follows:—

"*The Time required for Reflex Actions.**—When we stimulate one of our eyelids with a sharp electrical shock, both eyelids blink. Hence, if the length of time intervening between the stimulation of the right eyelid and the movement of the left eyelid be carefully measured, this will give the time required for the development of a reflex action. Exner found this to be from ·0662 to ·0578 sec., being less for the stronger stimulus. Deducting from these figures the time required for the passage of afferent and efferent impulses along the fifth and facial nerves, to and from the medulla, and for the latent period of the muscular contraction of the orbicularis, there would remain ·0555 to ·0471 sec. for the time consumed in the central operations of the reflex action. The calculations, however, necessary for this reduction, it need not be said, are open to

* Foster, *op. cit.*, p. 478.

sources of error. Exner found that when he used a visual stimulus, viz. a flash of light, the time was not only exceedingly prolonged, ·2168 sec., but very variable.

"The time required for any reflex act varies, according to Rosenthal, very considerably with the strength of the stimulus employed, being less for the strong stimuli, is greater in transverse than in longitudinal conduction, and is much increased by exhaustion of the cord. It has been stated that the central processes of a reflex action are propagated in the frog at the rate of about eight mètres a second; but this value cannot be depended on. The time thus occupied by purely reflex actions must not be confounded with the interval required for mental operations; of the latter we shall speak presently.

"*Inhibition of Reflex Action.*[*]—When the brain of a frog is removed, reflex actions are developed to a much greater degree than in the entire animal. We ourselves are conscious of being able by an effort of the will to stop reflex actions, such, for instance, as are induced by tickling. There must, therefore, be in the brain some mechanism or other for preventing the normal development of the spinal reflex actions. And we learn by experiment that stimulation of certain parts of the brain has a remarkable effect on reflex action. In a frog, from which the cerebral hemispheres only have been removed, the optic thalami, optic lobes, medulla oblongata, and spinal cord being left intact, a certain average time will be found to elapse

[*] Foster, *op. cit.*, p. 474.

between the dipping of the toe into very dilute sulphuric acid and the resulting withdrawal of the foot. If, however, the optic lobes or optic thalami be stimulated, as by putting a crystal of sodium chloride on them, it will be found, on repeating the experiment while these structures are still under the influence of the stimulation, that the time intervening between the action of the acid on the toe, and the withdrawal of the foot, is very much prolonged. That is to say, the stimulation of the optic lobes has caused impulses to descend to the cord, which have there so interfered with the action of the nerve-cells engaged in reflex action as greatly to retard the generation of reflex impulses; in other words, the stimulation of the optic lobes has inhibited the reflex action of the cord."

The effects of light have been referred to several times as producing marked effects upon the brain-centres, and exerting a stimulating or inhibitory influence. It may then be interesting to review a few of the known effects of light, and the absence of light, upon living things.

It is generally admitted that light does produce marked effects upon the human body. We see different phenomena occur in daylight and in darkness. Sunlight is most useful in the cure of anæmia, and partial darkness often helps the cure of an acute case of chorea; in megrim and in ophthalmia the patient is often unable to bear the effects of light. Some of the effects of the exclusion of light from cave-living animals have been well demonstrated by the author of the "Origin of Species" (see

p. 101). The action of light on vegetable organisms, and its effects in producing movements of various kinds, have also been carefully studied. I have, therefore, put together the following notes on the effects of light.

In such a subject as the present, as in so many other questions concerning the action of physical forces upon living organisms, it seems to me useful to consider the effects of light upon the simple cellular vegetable organisms, where these changes can be studied with much exactness, as well as in animals.

The action of light may be roughly described as trophic or kinetic; that is, mainly producing growth, or mainly causing movements and not growth.

Take a sample of the human species at the earliest period of possible observation, directly after birth. If we examine a new-born infant, we find that when a light shines upon its face it screws up its eyes (*i.e.* the orbiculares oculi muscles contract strongly), and it corrugates the forehead (*i.e.* corrugators contract). If the eyelids be opened, the iris contracts to light; in some cases, when the infant is a few days old, there seems to be a tendency to turn the head from the light, *i.e.* the skull rotates from a lateral light.

The action of light upon the iris needs only to be mentioned. It is commonly admitted that this is a purely reflex act in which the optic is the afferent nerve; the third, or motor-oculi, the efferent nerve; and the centre is some portion of the corpora quadrigemina.

The infant as born has, then, such an apparatus as allows of these two reflex actions to light. Now, in these reflex actions the light excites or stimulates these reflex movements. A plain beam of white light effects this. I do not know the effects of different rays of the spectrum.

I am not aware of other effects of light in the infant in its youngest stage, but, at five months old, holding a brightly coloured object well within the field of vision of the infant attracts its attention, as it is said. We see the child's eyes and head move, so as to bring their axes in a straight line towards the object. This is a mechanical effort; there is work done. What force can excite this movement and direct it with such precision? We know of no force passing from the object to the child, except the light reflected; a screen hiding the object from the child's eyes, *i.e.* cutting off the rays of light reflected from the object, prevents the object from exerting its effect upon the child. The beam of light reflected from the object excites the movement that follows action of the light; the effect of the light must penetrate the retina, and travel to the centres for the muscles seen excited to action.

A few words must now be said as to the effects of "the sight of objects." When we see an object, all our subjective impressions of that object are the effects of the beam of light reflected from the object; and all the objective, observable effects of the sight of that object are the effects of the excitation, or stimulation, produced by that beam of light. Here, again, the beam of light, or its effects, must be transmitted from the retina to the deep centres affected.

A lateral light falling upon the field of vision stimulates the opposite hemisphere, and this causes contraction of the head-rotatory muscles on the side from which the light comes, causing it to be turned towards the light or object seen.

Fig. 8.—Hydrophobia, after Sir C. Bell. Head repelled by sight of water.

Light acts as a stimulant of movement; conversely, darkness favours quietness. Strong light dispels sleep.

I think it can be shown that the stimulating effects of a beam of light reflected from an object may spread widely among the nerve-centres, producing co-ordinated and organized sequences of movements; this is the effect of the visual stimulus.

The sight of an object not only produces movements of the eyes and head, but also causes such movements in the upper extremity, as result in the hand approaching and grasping the object which reflects the beam of light and excites the movements; the movements spread to a wide area.

The sight of an object that has many times been seen, especially if some pleasure has accompanied the sight, produces a repetition of movements; the reflex action recurs on the recurrence of the same light stimulus. This indicates "retentiveness;" it is evidence of the fact that nerve-currents proceed along certain paths when they are produced by a similar stimulus.

In all these phenomena a moving beam of light produces more marked effect than a stationary beam; that is to say, a moving light, a moving object, attracts the most attention.

The brain of an imperfect subject may be so constructed as not to give the phenomena described above. The child may not be attracted by the light; it may, when six months old, not notice or attempt to grasp an object presented. We see that here the central nerve-mechanism is not duly susceptible to the reflex actions of light, and this indicates that it is not normal in its construction. These are some of the methods by which we judge whether a child has such a nerve-mechanism as will probably develop the due or normal functions of mentation in its after growth.

A further example of "retentiveness" to light impressions is seen in similar repetitions of facial expression, following the sight of the same object.

One effect of light, then, is to excite, stimulate, or stir up certain reflex actions.

A second effect of light is to produce new, not inborn, reflex actions, as indicated by the production of retentiveness in the nerve-mechanism to the effects of certain light impressions.

A third effect of light is to control or modify previously existing movements. Examples are seen in the effect of the sight of a bright light, or the sight of certain objects, in co-ordinating or arresting spontaneous movements.

A light, or the sight of an object, may inhibit movements instead of increasing their amount and frequency (see tracing, Fig. 9).

The following phenomena may then be observed as produced by the action of light on man :—

(1) Reflex action.
(2) Transmitted effects.
(3) Action on spontaneous movements $\begin{cases} (a) \text{ Excitant,} \\ (b) \text{ Inhibitory,} \\ (c) \text{ Co-ordinating.} \end{cases}$
(4) Retentiveness of the effects of light.
(5) The evolution or building up of mentation.

The action of light may be spoken of as extrinsic, inasmuch as it is a stimulus originating outside the organism, and producing changes in its structures. In contrast we may speak of the "passions" and "feelings" as intrinsic stimuli, as it may be supposed that they originate in some part of the organism.

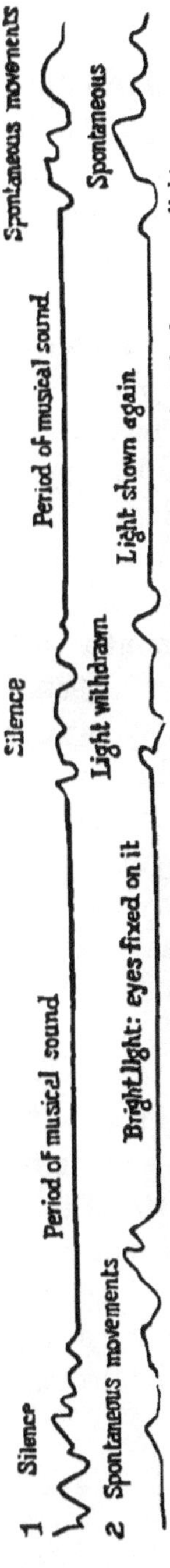

TROPHIC ACTION OF LIGHT. 101

Extrinsic stimuli may be—(1) immediate in action, as when a body acts by mechanical contact; (2) mediate, as when the sight of an object produces a visual stimulus which passes to the subject from the object seen. These mediate influences have sometimes been called sympathies. In the infant at birth, respiration is a mediate reflex act. Air excites the movement in a normally constructed child. Sucking is a reflex act due to immediate mechanical irritation, and is dependent upon the medulla oblongata. Such actions are called "instinctive." In this sense movements in plants are "instinctive," that is, due to their essential construction.

As to the more distinctly *trophic effects of light*. Pigmentation of the skin occurs in the tropical regions when light is strong.

Charles Darwin* gives an account of certain *cave-living animals* whose eyes have, in the course of successive generations, been lost owing to disuse. "It is well known that several animals, belonging to the most different classes, which inhabit the caves of Carinola and of Kentucky,

* "Origin of Species," p. 110.

Fig. 9.—Tracing, showing how the spontaneous movements of an infant were inhibited by sound and, again, by a strong light.

1 Silence
2 Spontaneous movements
Period of musical sound
Bright light: eyes fixed on it
Light withdrawn
Silence
Period of musical sound
Light shown again
Spontaneous
Spontaneous movements

are blind. In some of the crabs the foot-stalk for the eye remains, though the eye be gone; the stand for the telescope is there, though the telescope with its glasses has gone. As it is difficult to imagine that eyes, though useless, could be in any way injurious to animals living in darkness, their loss may be attributed to disuse. In one of the blind animals, namely, the cave-rat (*Wotama*), two of which were captured by Professor Sillman at about half a mile from the mouth of the cave, and therefore not in the profoundest depths, the eyes were lustrous and of large size; and these animals, as I am informed by Professor Sillman, after having been exposed for about a month to a graduated light, acquired a dim perception of objects."

This fact may, I suppose, be expressed by saying that in successive generations, owing to the want of the stimulation of light upon the parents' eyes, these organs have ceased to be developed in the young. It seems to me that this is equivalent to saying that the stimulating effects of light are necessary, in all probability, to prevent the organs specially suited to receive light impressions and convey them to the nerve-mechanism, from degenerating in successive generations. This is a trophic action of light.

The sight of a good dinner has been shown to increase the quantity of gastric secretion. Likewise the sight of food stimulates the salivary secretion. A woman in good health, after seeing her husband killed (light effect only), took up her infant and suckled it; the altered milk proved poisonous to the child.

The following table, after Sachs,* indicates the action of light on vegetation :—

A. General.

(1) Action of rays of different refrangibility.
(2) Variation in the action of light on plants in proportion to its intensity.
(3) Penetration of the rays of light into the plant.

B. Special.

(1) Chemical action of light on plants.
    (a) Formation of chlorophyll.
    (b) The decomposition of carbon dioxide.
    (c) Formation of starch in the chlorophyll.
(2) Mechanical action of light on plants.
    (d) The influence of light on the movements of protoplasm.
    (e) Cell division and growth.
    (f) Action of light on the tension of the tissue of the contractile organs of leaves endowed with motion.

Light may retard the growth of cells, as in heliotropism.

Darwin † has demonstrated many facts concerning the action of light on plants.

* Sachs, "Text-book of Botany."
† "Movements of Plants."

CHAPTER VII.

PATHOLOGICAL FACTS AND EXPRESSION IN PATHOLOGICAL STATES.

Disease may destroy or irritate parts of the brain—Destruction of corpus striatum—Lateral deviation of the head and eyes—Effects of irritation in contrast with destruction of a brain area—Effects of disease on different sets of muscles—Facial palsy—Localization of disease—Epilepsy—Chorea—Analogy to movements in plants—Experiments with the *Mimosa*—The study of chorea—Finger-twitching in nervous children—Tooth-grinding—Headaches in children; the physical signs—Cases of athetosis—Defects of development; their frequent coincidence.

THERE is no intention here to enter upon a description of conditions of disease from a medical point of view, but as direct experiments cannot be made upon the living brain of man, we avail ourselves of information derived from the rough kind of experiments often prepared for us when conditions of disease destroy or irritate certain brain areas; we thus observe the effect of destruction or direct irritation of parts of the brain.

In a case of complete hemiplegia, or paralysis of one side of the body, from a lesion destroying part of the corpus striatum of the opposite side, we find the following phenomena:—

Complete Hemiplegia from Lesion of the Right Corpus Striatum.*

1. The head turns to the right.
2. Both eyes turn to the right, and frequently both upper eyelids are fallen.
3. The muscles of the belly and chest are weakened on the left.
4. The muscles passing from the trunk to the left limbs are paralyzed.
5. The face is paralyzed on the left side.
6. The tongue on protrusion turns to the left.
7. The left leg is paralyzed.
8. The left arm is paralyzed.

Such is the combination of movements, and the compound posture, resulting from a destructive lesion of the right corpus striatum.

Now, to consider a few of the special movements in more detail. *Lateral deviation of the eyes and head*—both eyes turn to the right as a result of destruction of the right corpus striatum, and the head turns in the same direction. This is an interesting phenomenon for comparison with modes of expression by head rotation (see p. 185).

In the case of a right-sided brain lesion, the head rotates to the right on account of the weakened condition of the left muscles, so that those on the right side pull it over to the non-paralyzed side.

* See Dr. Hughlings Jackson in Russell's "System of Medicine," vol. ii. p. 537.

The eyes turn to the right as if looking at a person on the right side. Thus they do not lose their parallelism; the axes of the eyeballs remain parallel—there is no strabismus. This lateral deviation of the eyes may show the weakness of one side of the brain, and suggests a one-sided lesion in contradistinction to a general brain state, such as poisoning by opium. Now, observe the effect of such a lesion of the right corpus striatum (or of the centres which send their fibres through it) as causes convulsion of the parts previously paralyzed. In such a case we generally speak of an irritative lesion of the nerve-centre as producing the convulsion or spasm, in contradistinction to a destructive lesion which produces paralysis.

An irritative lesion of the right corpus striatum causes the head to rotate to the left, and the eyes to deviate to the left, while the left limbs are convulsed.

Right Corpus Striatum.

Destructive Lesion.	*Irritating Lesion.*
Head and eyes turn to the right. Left limbs paralyzed. That is, head and eyes turn away from the side paralyzed.	Head and eyes turn to the left. Left limbs convulsed. That is, head and eyes turn towards the side convulsed.

These facts help to give some explanation of certain coincident movements and postures (see p. 151). A destructive lesion on the right side of the brain weakens the left arm and leg, and the head rotates to the right. In an analogous manner we often see the left hand in the nervous posture—a sign of weakness of the right hemisphere of the brain; and this is often associated with rotation of the head to the right, with right inclination (see p. 151).

Again, we see that if one hemisphere, instead of being weakened, be excited, causing spasm of the muscles, the head is rotated to the same side as the spasm; thus the kind of condition of the brain is indicated or expressed by the relative position of the head to the side of the body affected.

The brain lesion causing hemiplegia does not affect equally all the parts of the upper extremity. On comparing the degree of power of movement in the fingers and in the arm, we find that though the patient can open and close his hand, and carry it to his mouth, still he cannot use the smaller parts of the hand for fine operations; he cannot pick up a pin off a wooden table, or unfasten the middle button of his waistcoat. He has lost the power for fine adjustments of the small parts, but has still fair power over the limb in its larger parts.

As to the facial paralysis seen in hemiplegia,* all the muscles on the side of the face are weakened, but very unequally; there is very slight weakening of the orbicularis palpebrarum. The patient can close his eyes, although not so strongly on the paralyzed side when urged to close them both tightly; sometimes, especially in chronic cases, we discover no difference on the two sides. This cerebral-facial palsy differs from that due to disease of the facial nerve in its distribution. Brain facial palsy weakens mostly the muscles about the mouth. This is seen markedly when the patient shows his

* See Figs. 10, 11, and compare with Fig. 27, p. 202.

teeth, or whistles; the groove running from the nose to the mouth is less marked than on the other side.

In studying any case of brain disease, it is very desirable to localize the seat of lesion. The following are briefly the principles employed in trying to make such localization during the life of the patient. It is impossible to make any scientific

Fig. 10.—Right Hemiplegia, with cerebral facial palsy, right side. The face is asymmetrical, and the muscles in the right lower zone about the mouth act very indifferently. The naso-labial groove on this side is almost lost; this is well seen on comparing the two sides. No asymmetry is seen in the upper and middle facial zones.

diagnosis of the locality of the lesion unless some localizing symptoms be present. Now, the main localizing symptoms are paralyses and spasms of the muscles supplied with motor force from the nerve-centres affected by the lesion. Destruction of the corresponding mass of brain substance, in different men, whether by hæmorrhage or by softening, paralyzes the same muscles in each, and interferes with corresponding movements; hence, when we find one particular set of movements interfered with, we infer the seat of lesion.

Similarly, convulsion or spasm in one particular set of muscles indicates the discharge of motor force from the particular set of nerve-centres whose destruction leads to paralysis of that particular

Fig. 11.—Left Hemiplegia, with cerebral palsy, left side.

set of muscles. The irritation causing such a discharge of force may be a tumour, or local inflammation irritating that portion of brain. Hence coarse or extensive paralyses, and other profound disturbances of the nerve-muscular system, have received much attention from clinical and pathological observers, and by the accumulation of their joint observations much knowledge has been gained as to the symptoms that result from lesion of certain portions of the brain. This encourages us to observe, in all cases of health and disease, the movements and results of movement, knowing that these correspond to, and are the direct outcome of, the states of certain nerve-centres. The knowledge that we already possess of the nerve-centres is from observation of muscular action. In a given case, by comparing the state of the muscles during life, as they may be affected by paralysis or spasm, with the brain lesion found after death, and by collecting and comparing many cases, it has been found that destructive or irritative lesions of certain parts of the brain cause paralysis or spasm of a certain set of muscles corresponding.

Epilepsy is a chronic disease, characterized by (1) attacks of more or less disturbance of consciousness, (2) muscular spasms and convulsions. The careful, detailed, and accurate observation and record of the movements of the convulsion, and sequent paralyses, has proved an efficient means of study in this disease, the movements and results of movements being studied as indices of the action of nerve-centres. Epilepsy is a condition of disease of which we know but little beyond what can be learnt from studying nerve-muscular movements, and their associations and concomitants; still we do possess much practical and useful knowledge of epilepsy.

Chorea is a disease, or abnormal condition, commonly seen in children. It is characterized by a great excess of involuntary movement, and a varying amount of muscular weakness. The movements cease during sleep; in kind they resemble gesticulations, and in their combinations and successions are probably such as may occur during conditions of health. Each movement probably depends upon a discharge of motor force from some nerve-centre corresponding. Chorea is dependent upon a brain condition which we know only through the effect produced upon the muscles by the brain.

In children I have often observed that "the weak and nervous" have much spontaneous finger-twitching; and I described this as one of the physical signs seen in children who suffer from recurrent headaches and associated pathological conditions.* Such muscular unsteadiness seems very analogous to the movement of young, growing, sensitive vegetables.

The two tracings (Figs. 12, 13) indicate the continuous condition of spontaneous muscular unsteadiness of the finger of a nervous child; and the continuous involuntary movement appears analogous to that indicated by the tracing of the movements of some plants. Now, if this analogy between unstable mobile vegetable cells, and unstable nerve-cells, be legitimate, it should guide us to further useful observations.

To be brief, Darwin's observations show that movement produced by the growth of vegetable cells is constant in the leaves, stems, and roots of many young plants.† If the movement of nervous

* *British Medical Journal*, December 6, 1879; see also "Brain," 1881, parts xi., xii., xiv.
† See the account given of circumnutation in plants, chap. ii.

children be produced by a condition of brain-cells analogous to that of the growing parts of plants or the cells of the pulvinus, it should be liable at times, under certain circumstances, to great exacerbations. Thus guided, I have taken tracings of the finger-movements of nervous children and of those suffering from chorea. Samples are presented here, and seem to indicate the following results:—

1. The movements of chorea are far more fre-

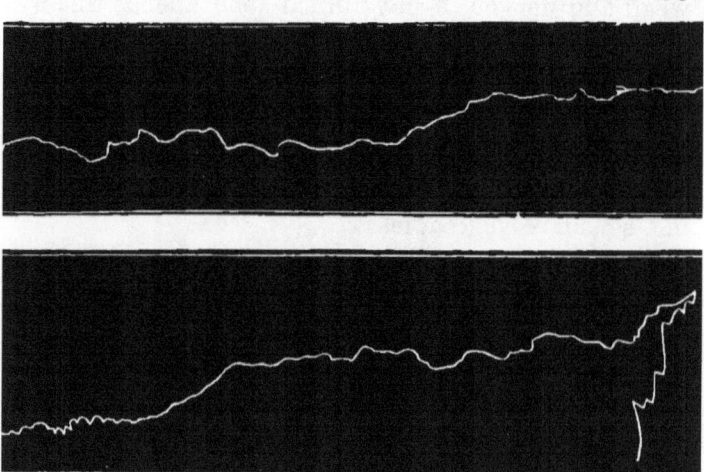

Figs. 12, 13.—Tracing of involuntary movements of the finger in a nervous child.

quent and continuous than might be expected from mere inspection of the hands.

2. These movements may be but an exaggeration of the movements of a nervous child, usually present, but often overlooked, tracings of which are given in Figs. 12 and 13.

3. The twitching movements of chorea may be compound, each visible twitching being compounded of many of the little movements seen in the other

tracings. I have never found such compound tracings in tremors as in paralysis agitans.

To prove these points with certainty, numerous tracings from many cases would have to be compared.

As to the treatment of the class of children referred to, the following experiments are very suggestive, and the analogy to the case of children

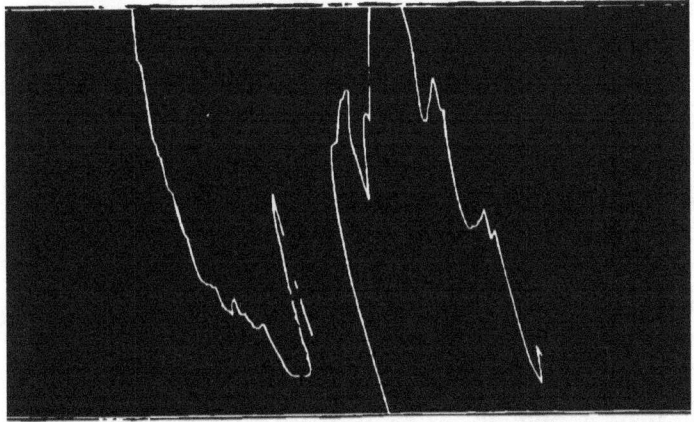

Fig. 14.—Finger-tracings in chorea. The twitching movements are compound.

hardly appears to require verbal description. Mr. M. Voss, of Streatham, has kindly communicated to me the results of his investigations.

Three years ago, some seed of the sensitive plant (*Mimosa pudica*) was set to grow, and at a moist heat of about 90° Fahr. it soon germinated. Before the compound foliage growth had commenced, the seedlings were potted off into different earths and sand. Those planted in a soil of two parts of decayed vegetable mould to one of sand grew more vigorously both in height and foliage than the others; and, after two months' growth, they were much less sensitive than others planted in two-thirds of silver sand and only one-third of leaf-mould. One or two plants were grown entirely in silver sand. These showed extreme sensitiveness to the slightest touch; even a breath of air, or the slightest jerk of the pot in which they grew, caused all the foliage to shut up. Those plants having no nourishment beyond the gases in the air, or sand, soon turned yellow and died. The plants in two-thirds sand and one-third decayed vegetable mould were not so robust or strong as those grown in a greater proportion of vegetable mould. They failed to produce any flowers, and died off at the lower temperature to which all the plants were exposed, whilst those planted in two-thirds vegetable mould and one-third sand, fully matured their growth, flowering in a temperature of 50° or 60° Fahr., the foliage being of that full green colour denoting the fact that the spongioles of the roots had necessarily been supplied with the various chemical gases in the soil (set free by a due amount of moisture) requisite for producing the continued support of the plants. Their sensitiveness had, at the end of August, almost left them; indeed, after a blow on the leaf with a twig, the foliage would fall, but almost immediately regain its horizontal position.

Many other useful analogies might, I think, be made in this direction; and the subject is one full of interest to all who look upon the conditions of health or disturbance of the system throughout the whole organized world. Among other matters, we might consider the transmission of sensitive impressions from one part of the plant to another; the metamorphosis of parts of a plant according to the functions discharged; the irritability or sensitiveness of plants, which is in some cases far greater than anything seen in man; the production of acid secretions in mobile plants; and numerous other facts easily observed in plants — facts of great interest to the physiologist and pathologist; and in this direction Mr. Darwin's researches, following those of Sachs, are a climax to investigations carried on since the time of Sprengel; facts which Sir James Paget * would have us apply to the study of human pathology.

In different cases of chorea very different groups of muscles may be affected, thus indicating the very different brain areas that may be choreic. In studying a case of chorea we try and indicate the extent of brain affected by specially indicating the choreic area. The following points then should always be looked for—as present or absent in any case described—and the order of invasion of groups of muscles or their recovery should be observed:—

1. In examining a case to prove the fact of chorea it is very convenient first to look at one or both hands, held out free and disengaged. The kind of movements of the hand and fingers have been dwelt upon and described.

2. The upper and lower extremities present the greatest mass of the choreic movements. It is

* See his Lectures, *British Medical Journal*, October 16, 1880.

important to note whether the finer or coarser movements be the most affected; the amount of involuntary movement, and the power of voluntary act that is left.

3. Hemiplegic varieties are common; the least mobile side may be much weakened, though not much moved.

4. The face. Varieties in this group of muscles have been discussed.

5. The soft palate may present marked movements of an irregular twitching kind, the levator-palati muscles working distinctly. I do not refer here to the dragging of the palate by the choreic twitchings of the tongue, but to the primary twitching of the palatine muscles. In some cases the levators are distinctly seen twitching upwards. This symptom is often absent in chorea, and when seen, I have observed that it has usually passed off early.

6. The tongue may be jerked in and out. When protruded, it may present much movement, but still be kept out a fairly long time.

7. Eyes. Upper eyelids often strongly retracted. Eyeballs often much moved.

8. The head in the active stage is often moved much. During convalescence, and when the active movements have passed off, a lolling of the head to one side is common; *i.e.* inclination with rotation to the same side, combined with slight flexion.

9. The spinal muscles and trunk are often affected. The child often balances itself very ill, throwing the scapular and upper dorsal region too far back,

and thrusting the pelvis too forward, the spine still remaining symmetrical, or being thrown into lateral curves. It has seemed to me that the awkward appearance was due to want of adaptability of the proper compensations in the movements of different muscles.

10. The respiratory muscles. These may be affected much or little; the alæ nasi muscles may be affected also.

11. The vascular centres. The heart's action is sometimes irregular.

After looking at cases of chorea we naturally consider other nervous cases characterized by nervemuscular movements, and if we study children we find examples abundant.

In 1879 I* put together my notes of fifty-eight cases of children of nervous mobile temperament; and for this practical purpose a summary has been made of the principal symptoms of cases of headache in children under fifteen years of age. Though recurrent headache was the most prominent and constant symptom met with in this group of cases, it was not always the one complained of by the patient or mother. Classifying the symptoms, we find the largest number referable to the mental and cerebral condition of the child. We are told that he is excitable, melancholy, passionate, or fretful; that sleep is restless and disturbed by dreams and night-talking, by attacks of terror and screaming, or by somnambulism. Often there are vague pains in the limbs, chest, or abdomen. The

* See "Brain."

appetite is often variable, voracious or entirely lost; there may be considerable emaciation with a good appetite, and without any accompanying signs of organic disease; often there is considerable nausea or retching with the attacks of headache. Hacking cough, without physical signs to account for it, is often met with, and usually the child is excessively fidgety.

In studying this group of nervous excitable children in whom recurrent headache was a main symptom, it was first necessary to look out for certain objective signs by which they might be indentified; and these were principally found in the condition of the nerve-muscular system, the teeth, and the urine, but chiefly in the muscular system.

The nerve-muscular condition was carefully observed in this group of cases. Usually, the child was made to stand with the heels together, looking straight before him, the hands being down by his side. The general steadiness of the body, trunk, and limbs was then observed, as well as the condition of the face and eyes. He was next directed to put out his tongue, and, finally, to hold out both arms straight before him, on a level with the shoulder, at the same time separating the fingers. Observations were then made as to the muscles of these parts. The state of the heart, lungs, teeth, and urine was examined, and the patient's weight was recorded. As the result of these examinations, it was observed that frequently there was marked fidgetiness, and irregular movement of the trunk and limbs; the trunk was often swayed about—

frequently it was inclined backwards, apparently to preserve equilibrium, while the hands were held out in front. Twitchings of the fingers * were very commonly seen, the movements being usually of independent digits, and most commonly either lateral or flexor, less commonly extensor; the lateral movements appeared to be the most characteristic. The tongue was usually very unsteady, but not distinctly jerked in and out, as in chorea. Twitchings of the facial muscles were not very common.

A certain passive condition of the hand and face will be referred to presently. In a large proportion of the cases, the teeth were found flattened on their edges, as the result of "tooth-grinding;" the teeth mostly ground were the incisors and canines, but the special teeth flattened depended upon their arrangement in the jaws. The pupils were measured in some cases with a catheter-gauge, after the manner suggested by Mr. Hutchinson. No special conclusion was arrived at; on the average, they were not large, except in a few cases near the age of puberty.

The nervous hand was found to be very common; the expression of headache in the face was frequently seen, or rather, I should say, a certain facial expression indicative of depression was often observed, and, when seen, appeared to give strong evidence, characterizing the child as the subject of recurrent headaches. To analyze this condition of the face, the faces of adults, the subjects of migraine, were studied. The most noticeable

* See Figs. 12, 13.

point was the look of depression, and heaviness and fullness about the eyes, especially the under eyelid. If a paper were held so as to cover either half of the face, the expression observed still remained, proving the condition of the face bilateral; if the forehead above the eyebrows were covered, or the face below the lower margin of the orbit, in each case the expression seemed still apparent; while, if the paper were held so as to cover that portion of the face which lies between the eyebrows and the lower margin of the orbit, it seemed impossible to recognize the peculiar *facies* under consideration. It appears that this expression must be due principally to the condition of the orbicularis palpebrarum. Specially observing this muscle and the parts adjacent, there seemed to be a loss of tone in the muscle; there was an appearance of fullness and flabbiness about the lower eyelid; the skin hung too loose, with an increase in the number of folds; and, in place of falling against the lower eyelid neatly, as a convex surface, it fell more or less in a plane from the ciliary margin to the lower margin of the orbit, a condition that is often best seen by looking at the patient's face in profile. This condition of the parts about the eye was unaccompanied by any general change in the skin of the face, such as the flabbiness seen in emphysema, and the loose inelastic skin of senile decay; further, the facial expression is not at all necessarily permanent, but may disappear with improving health, and it is removed if the patient can be made to laugh. It is not suggested that this muscular condition only accompanies headache; it appears common to other conditions of depression.

Having attempted to demonstrate that we may ascertain something of the condition of the brain in these nervous children by observing these muscular movements, we may now look for signs of irritation of the cranial nerves. Evidence of disturbance of the motor division of the *fifth nerve* is seen in the great frequency of tooth-grinding; the condition of the muscles supplied by the facial and hypoglossal nerves has been referred to. Irritation of the pneumogastric nerve appears to be indicated by many symptoms. The varying appetite, which is often voracious, though nutrition is deficient, while at other times it is markedly defective, the frequent epigastric pain, and the retching or vomiting with headache, appear to indicate disturbance of the gastric branches; occasional palpitation without heart-disease, and the frequent hacking cough without signs of lung or throat mischief, indicate probable irritation of the cardiac and respiratory branches.

As to the disturbance of sensory nerves, it was difficult to obtain evidence, as the patients were often unable to describe their sensations with accuracy. In five cases, varying from nine to fourteen years of age, distinct dysæsthesia of vision accompanied the attacks of headache, the patient seeing colours, sparks, or other illusions during the attack of headache; in all but one case, the mother also suffered such spectra with headache.

Tooth-grinding is produced by the action of the deeply situated pterygoid muscles; champing of the jaws is produced by the masseter and temporal muscles; all these muscles are supplied by the fifth nerve, and it is to their condition that we must look for information as to the condition of the central origin of the nerve. Tooth-grinding, when it has become a habit, is indicated by the flattened condition of the tips or edges of the teeth, which may be ground down—a sign that may be particularly seen in the incisors and canines. Ground teeth are very common in nervous children, such as those who suffer from recurrent headaches, restless sleep, somnambulism, and finger-twitching. In lunatic asylums and wards for imbeciles it is very common to hear tooth-grinding on every side; in such cases tooth-grinding is a sign of central irritation of the fifth nerve. It is well to bear in mind that the sensory branches of this nerve supply the membranes of the brain and the external parts of the head.

Seeing that the slight disturbances occurring during sleep in many children causes the pterygoids to contract rhythmically, it is not surprising that grave disease should cause spasm of the other muscles supplied by the fifth nerve, as in epilepsy and hysteria.

The ninth nerve is motor to the tongue, and this organ being a mass of muscular fibres running in various directions almost unsupported by bones, is very sensitive to changes in the nerve-centres. In chorea the tongue is often jerked in and out in a manner quite characteristic of the disease; in other cases it is easily kept protruded, and its substance is seen to be in a condition of constant movement.

Such irregular movement is also very common in nervous children; a tremulous tongue is characteristic of alcoholism, and general paralysis of the insane.

As to the condition of the mental centres, disturbed and restless sleep was very common; night-terrors were frequent; often the child would scream out that "a lot of people were coming to kill him," that he "saw the school-board man" coming, etc. In six cases, there was the distinct history of somnambulism; in four of these, the acts performed during sleep had been complicated and curious; in one case, that of a boy nine years old, such attacks were frequent at the time he came under observation, on account of his headaches; in the remaining cases, somnambulism had occurred at an earlier period.

In seeking the lines of causation of recurrent headaches in children, the cases were arranged in a tabular form according to age and sex.

Ages		3-4	4-5	5-6	6-7	7-8	8-9	9-10	10-11	11-12	12-13	13-15
Males	25	1	2	2	8	2	1	2	2	4	1	0
Females	33	0	2	3	1	2	5	5	4	2	4	5
Totals	58	1	4	5	9	4	6	7	6	6	5	5

As among other groups of nervous cases, the preponderance of number is with the female sex. Heredity appeared to produce a marked predisposition to this neurotic condition. There was a history of recurrent headaches in the mother in twenty-four cases, and in the father in eight, while in three cases there were examples of insanity in the family.

As to treatment, the restless, excitable condition of these children, and the great want of rest in sleep, appeared to indicate the use of bromides and other sedatives; and this plan of treatment was generally adopted, tonics being occasionally used, together with small doses of chloral at night for short periods, till the habit of sleep was induced. Under treatment, marked improvement occurred in many cases, the child gaining one or two pounds in weight in a month or six weeks, at the same time losing the headaches, sleeping quietly at night, and again becoming fit for a child's school-life.

Dr. Hughlings Jackson (*Lancet*, July 11th, 1875) has shown that, in cases of chorea, paroxysmal headaches are of common occurrence. Dr. Herman collected the histories of seventy-six cases of chorea, and found that paroxysmal headaches occurred in fifty-three cases; in these, the headaches were not often preceded by ocular spectra. In the group of cases described in this paper, certain active and passive conditions of the muscles, especially the small muscles, are commonly met with in conjunction with disturbance of the higher nerve-centres; this appears to afford evidence that the two conditions may be owing to disturbance of the same nerve-centres, possibly the same as are affected in a greater degree in chorea. A girl, thirteen years of age, came under observation, complaining of frequent headaches, with ocular spectra, restless nights, tooth-grinding, and slight muscular twitchings, such as have been described. She had never suffered from rheumatism, but the family were rheumatic; she had no cardiac disease. Three years previously, she had had acute general chorea for four months; it is submitted that probably the same nerve-centres were affected in the greater and lesser illness, but in a different degree.

The varieties of finger-twitching have been referred to. They may be described as—

(1) *Flexor-extensor;* the primary movement being that of flexion, followed by a secondary extensor movement. This may be seen in a variety of cases, and in particular is seen in what is called "picking the bedclothes" in the typhoid state preceding fully developed coma.

(2) *Extensor-flexor;* the primary movement being that of extension, followed by a secondary flexor movement. This is common in the slighter forms of chorea and in nervous children; such twitches usually constitute the subsultus tendinum so indicative of exhaustion in the course of typhoid fever.

(3) *Abductor-adductor* twitches; the movements consisting in lateral separations of the fingers, followed by their being drawn together again.

As to "finger-twitching," the "nervous hand," and "the relaxed orbicularis oculi," the following statistics are from an analysis of thirty-four cases from my note-books of the East London Children's Hospital:—

Finger-twitchings in 19.—Twitchings alone . . in 8 cases.
With the nervous hand „ 8 „
With relaxed orbicularis „ 3 „
                                                     ——
                                                     19 „

The nervous hand in 19—Nervous hand alone . „ 7 „
With twitchings . . „ 8 „
With relaxed orbicularis „ 4 „
                                                     ——
                                                     19 „

Orbicularus relaxed in 10.—Orbicularis relaxed alone in 3 cases.
With the nervous hand „ 4 „
With twitchings . . „ 3 „
—
10 „

As to the general character of this group of thirty-four nervous cases in which nerve-muscular signs were specially noted, no cases of known organic brain disease were included, and all were under fifteen years of age. I have abstracted and summarized the diagnosis of the nineteen cases in which the "nervous hand" was seen :—" Headaches," 6 ; " neurotic temperament," 3 ; " anæmia and headaches," 2 ; " headache and somnambulism," 1; "restless sleep," 1; "laryngismus," 1; " a dull child with congenital ptosis," 1 ; " old rickets," 1 ; " debility," 2 ; " slight chorea," 1. The cases of " finger-twitching " had the same general characters as those with the " nervous hand," therefore I do not further describe them.

As to the kinds of finger-twitching, the varieties were noted as follows :—Simple twitching, 9 ; flexor and adductor, 5; flexor, 3; abductor-adductor, 1; extensor and abductor-adductor, 1. As shown in the tables above, the "nervous hand " was associated with " finger-twitches " in eight cases.

In cases where the right and left hands were compared, we find a difference in six cases, always to the disadvantage of the left hand ; it specially presented "the nervous position" in four cases, and finger-twitches were specially marked on the left side in two cases.

The cases with "relaxed orbicularis" were specially marked by recurrent headaches, some with optical illusions and scarlet zigzag forms.

In two cases herpes zoster occurred while under observation.

Athetosis is a condition of brain disease characterized during the life of the patient by movements, more or less constant, in the muscles of the extremities. Athetosis usually affects one side of the body only, but may affect both sides. The movements are much slower than those of chorea, and are not usually a series of gesticulations, but differ in their combinations from the movements performed in health. The plate gives the common postures seen in their different cases; they are postures that would not be assumed under the guidance of a normal brain influenced by emotions.

To give some idea of the expression of this condition of brain disease, I quote the account of the movements seen in three cases which I published.*

CASE I.—*Athetosis with Epilepsy.*

The girl was of fair stature, rather pale and thin, and of weak intellectual power. She could walk, assist her mother in the house, and answer questions, but was very dull and apathetic.

When seen the following description was taken: —*The left upper extremity* is almost useless for most voluntary purposes on account of the condition of the hand. If told to put her hand to her mouth, or to the back of her head, the hand gets there, being moved slowly and awkwardly by the action of the muscles moving the shoulder and elbow, but though flexion and extension of the elbow be suffi-

* "Brain," part xiii.

PHYSICAL EXPRESSION.

Fig. 16.—Tracings of Movements in Athetosis.

ciently voluntary to perform such acts, involuntary movements of pronation occur, while the hand is being moved up to the mouth. All through the time of the general movement of this limb it is obvious that there is much perfectly involuntary movement of the wrist, often causing marked flexion, or affecting independent digits.

As to the condition of the hand, the description was taken while the forearm was gently supported by the mother, the wrist being left perfectly free. It was then noted: The wrist droops, the metacarpus hanging straight down, but frequently supinated rather quickly, but not in a jerky manner; the hand then passes slowly back to its former position, the series of movements being repeated. Flexion and extension of the wrist through a few degrees are nearly constant movements.

The movements of the fingers are almost incessant, the primary and quickest movements being flexor, followed by slower extension. The movements are slower than the jerks of chorea, and more deliberate, not looking like gesticulations; they are purposeless and gliding in kind.

The fingers are constantly going through grotesque movements, sometimes crossing one another, sometimes one is extended, while the others are flexed; there are no adductor and abductor twitches, such as are usually seen in chorea; the thumb is mostly turned in on the palm, but is also often extended. These finger movements appear utterly purposeless, looking as if the tendons were merely pulling upon dead fingers, as an automaton pulled by strings, or like the movements of a boneless limb, such as an elephant's trunk or an anemone's tentacle. The metacarpal bones are rather contracted together by the arching of the palm of the hand, giving it a narrow appearance.

CASE II.—*Athetosis associated with Chronic Hydrocephalus.*

His general health was good, and no disease was detected in any organ but the brain. The condition

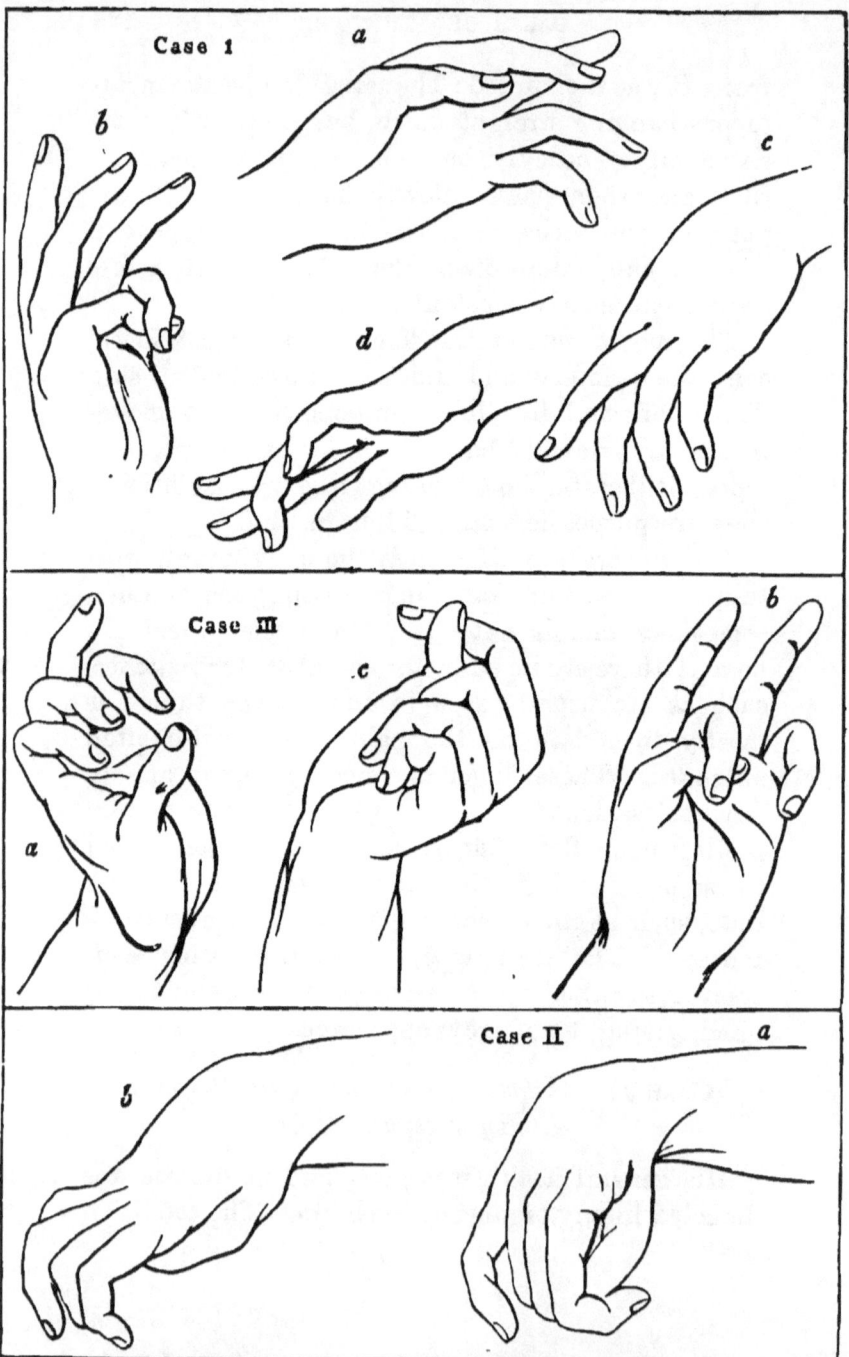

Fig. 16.—Cases of Athetosis showing Hand Postures.

of the right upper extremity especially attracted attention. The following description is compiled from notes taken on various occasions:—

The limb was useless for any purposive movements; there was no voluntary power over the hand, and but little over the shoulder and elbow. When a flower was held up to the child, he looked at it and made a noise, evidently indicating his pleasure, then moved his left hand to the flower, took hold of it, and tried to put it to his mouth. He could not take the flower with his right hand.

The wrist was frequently kept half flexed, while the fingers were extended and flexed generally altogether, in a slow and rhythmical manner. Pronation and supination were not common; the movements of the wrist were usually confined to flexion and extension of the fingers as described.

The hand was moved when pricked. There was no difference between the similar measurements in the two arms, but there was some rigidity of the right arm. The right leg was weak, but presented no athetosis.

Here the condition of the limb was associated with brain disease and convulsions. The movements occurred in a paralyzed arm, and were more limited in extent than in Case I.

CASE III.—*Double Athetosis not associated with Epilepsy.*

Eliza R., aged four and a half years. She was a playful, intelligent, pleasing little girl, good and well-behaved, and soon became a general favourite.

Her general health was good, and development appeared good in all particulars, but for the athetosis; she was well nourished.

The child was unable to stand or do anything for herself on account of the movements of the hands and feet, associated with which there was, no doubt, some want of muscular power. The muscular condition of the head and face appeared natural. There appeared to be a fair amount of voluntary power over the shoulders and elbows, so that she could hold a doll to her. When told to take hold of a toy she brings her hand to it, but is unable to open the fingers so as to clasp it; all through the time of this effort the fingers are in constant movement. When the object is placed in her hand she clutches it and is unable to drop it.

Supporting the left upper extremity free from the trunk by holding the humerus, it was possible to observe the following movements succeeding in a rhythmical manner.

The wrist was frequently bent backwards, and sometimes abducted. The thumb is mostly bent in on the palm, turned in under the index finger. The fingers are frequently extended at the meta-carpo-phalangeal joints, remaining flexed at the inter-phalangeal joints. The greatest power of extension seems to be of the index finger. In getting the hand near an object it moves about much before it comes in contact with the object, and then cannot grasp it on account of the condition of the fingers. The fingers are usually kept flexed; she cannot spread them voluntarily. When the child is played with, her fingers spread open spontaneously, as also when attention is specially drawn to the other hand. When one's finger is slipped into her hand, her fingers grasp it, and cannot let it go unless the fingers happen to open of themselves. When she makes an effort and puts her legs out of bed to show her feet, the involuntary movements of the hands and fingers are increased, and movements of pronation and supination are noticeable.

The toes of both feet are continually being flexed and extended, but these movements are less in degree, and less characteristic in kind than those of the hands. She can kick her legs about in bed, but cannot walk.

The eyes and special senses are normal, and general sensation is good. Heart and lungs healthy.

The movements are very strange and purposeless; in kind they have more resemblance to voluntary movements than is usual in athetosis, and less of the gliding, successive-rhythmical character.

During sleep the hands are usually quiet.

This condition of the child appeared to have existed from birth. There were no signs of progressive disease.

The family were quite healthy.

Defective Developmental Conditions as seen principally in Children.[*]

In seeking for explanation of the circumstances attendant upon and causing some of the special

[*] From *Medical Times and Gazette*, January 21, January 28, and February 11, 1882.

developmental conditions commonly found in children, it seems necessary, first, to arrange and examine cases of gross and obvious deformities, where the kind of ill-developments, and any co-existing defects and consequent deviations from normal function, may be easily observed and recorded. For this purpose I have abstracted from my case-books the notes of twenty-three cases, imbeciles and idiots being generally passed over. These are arranged as follows :—

Group I.

Case 1.—Heart-defect—Fingers clubbed—Cyanosis—Palate cleft.

Case 2.—Heart-defect—No cyanosis—Deformity of hands—Epilepsy in family.

Case 3.—Mitral and tricuspid disease—No cyanosis—Malformation of hands.

Case 4.—Heart-defect—Congenital cyanosis—Left hemiplegia, dependent upon defect of right hemisphere—Bell's paralysis on right side of face, with deformity of right ear.

Case 5.—Heart-defect without cardiac symptoms—No cyanosis—One ear deformed.

Case 6.—Congenital heart-defect, with a varying bruit—No cyanosis—Patency of inter-auricular septum—Want of power in legs from birth, with some rigidity of left.

Remarks on Group I.—This series of six cases illustrates the concurrence of congenital defect of the heart with other deformities, *e.g.* cleft-palate, defects of hands, ill-formed ears, congenital defect of brain. Of this latter many examples might be added, and I propose at another time to give more examples of concurrent defects of heart and nerve-centres. In some of these patients the heart was specially examined, not on account of any signs of heart-defect, but in the search for examples of concurrent congenital defects. Œdema was absent in all these cases. Cyanosis was present in Cases 1 and 4, and absent in the other four cases of this group.

Group II.

Cases with evidence of congenital heart-defect, not associated with other known deformities.

Case 7.—Heart-defect—Cardiac hypertrophy—No bruit—Marked cyanosis—Convulsions—No other defects.

Case 8.—Heart-defect—No cyanosis—Bruit in pulmonary area—No other defects.

Case 9.—Heart-defect—Cyanosis—Bulbous fingers—No other deformities.

Case 10.—Heart-defect—Cyanosis—No clubbing of fingers or toes—No other deformity.

Case 11.—Heart-defect—Varying amount of cyanosis—No bruit—No other deformities.

Remarks on Groups I. and II.—Of these eleven cases of heart-defect no co-existing deformity was found in five. Cyanosis was present in six cases out of the eleven; and it is noteworthy that of these cyanotic cases the larger proportion were in Group II., where no co-existing deformities were found. In Case 10 cyanosis was more marked in the feet than in the hands. As to the signs of the presence of heart-disease, bruits were present in nine cases; hypertrophy, evidenced by forcible impulse, distinct area of dulness, strong pulse, or proven by autopsy, was present in seven cases; irregularity of action was present in three children. Clubbing of the fingers in three cases was noticed. No symptoms dependent upon the heart-defect were noted in six of the cases given. In six cases some kind of evidence was obtainable as to causation. I found no evidence of "maternal impression" as a cause of congenital defects of the heart; but there was evidence against the family in four cases in such particulars as to many deaths or several miscarriages preceding the birth of the patient. In two cases there was no bruit, and in another case the presence of a bruit was very doubtful.

Group III.

CASES OF CLEFT-PALATE.

Case 12.—Cleft-palate—No heart-defect—Head small.

Case 13.—Cleft-palate—No heart-defect—Head small—Premature birth—Marasmus.

Remarks on Group III.—Neither of these cases presented any known coexisting deformity, but in Case 1 a heart-defect accompanied cleft-palate. Such a coincidence appears not improbable, and is worth looking for in other cleft-palate patients as a question affecting prognosis and the safety of chloroform in operating, etc.

Group IV.

DEFORMED UPPER EXTREMITY, AND INTRA-UTERINE AMPUTATIONS.

Case 14.—Fingers webbed—Ears not symmetrical—Excessive epicanthic fold *—Cerebral deficiency.

Case 15.—Deformed hand—Intra-uterine amputation (?)—No other defect.

Case 16.—Deformed arm—Intra-uterine amputation (?)—No other defect.

Remarks on Group IV.—Two of these cases presenting defective upper extremities looked as if they had been intra-uterine amputations, and being traumatic, and accidental rather than developmental, it is not surprising that no other defect coexisted. In Case 14 we see coexisting defects. The asymmetry of the ears and the webbed fingers were surely developmental defects, and they were accompanied by defective development of hands.

Group V.

MISCELLANEOUS CASES.

Case 17.—Double coloboma of iris—No other defect.

Case 18.—Congenital smallness of one eye—No other defect.

Case 19.—Deformity of right ear and temporal bone—No other defect—Rickets.

---

* The term "epicanthic" fold is applied to the portion of skin that lies at the inner angle of the opening of the eyes towards the nose. This is occasionally developed as a kind of web, called the epicanthic fold.

Case 20.—Congenital jaundice—Double hydrocele.
Case 21.—Ichthyosis—No other deformity.
Case 22.—Ichthyosis, moderate in degree—No other defect.
Case 23.—Ichthyosis—Deformity of both ears—Heart healthy.

Of the twenty-three cases, thirteen were males and ten females. Looking at these cases from the point of view indicated, the following points seem worthy of consideration :—The coexistence of deformities was not uncommon. This is seen in all the cases in Group I., also in Cases 14 and 20; while in Nos. 12 and 13 the head was also below the average size. The family history is noteworthy as giving indications of possible causation. In Cases 2, 6, 10, and 14, a feeble constitution in the family is indicated by miscarriages, many deaths, insanity with epilepsy in previous members of the families before the birth of the patients described. It also appears in some histories that the tendency to ill-development exhausted itself, the later members of the family appearing healthy. The secondary effects of the congenital defect are important. Defect of heart may lead to cyanosis, clubbed fingers, and perhaps low temperature; it is also said to lead to a low mental development. If the internal ears are faulty, dumbness may follow; obstruction of the common bile-duct must secondarily cause jaundice. Ichthyosis being attended with inaction of the skin, secondary bronchitis is common. Cleft-palate may lead to atrophy from inanition, and a head below size may lead to general organic feebleness; but the former condition may be rectified, and not lead to atrophy; in a former paper (*British Medical Journal*, October 30, 1880) I showed that small-headed children may, under proper care, develop a fairly sized brain. Here the developmental defect is, to some extent, removed. Of the eleven cases of heart-defect only one appears to have had convulsions—a symptom constantly inquired for. This case (No. 7) was cyanotic; there were, however, six cases of cyanosis without any history of convulsions.

It will probably be admitted that in the structure and general anatomy of the human body "the normal" is but the average as it is found; and so difficult is it in all the organized world to define and distinguish accurately between "the variation and the monster," either in the seedling plant which is different from its ancestors, or in the "genius" in the human species, that we must look carefully before we say that any specimen is monstrous or diseased; but in the gross cases above narrated the abnormality is obvious. Possibly, in these co-existing defects of development we may see some explanation of the accompaniment of vulgar faces and low minds, *i.e.* low development of the brain; when such samples are seen in a family it may be well to look for the lines of causation in the descending scale of the development of the family. Among common defects may be enumerated defective or excessive ossification of the skull, excess of the epicanthic fold, defects of the eyeball, webbing of fingers and toes, nævus, etc.

CHAPTER VIII.

POSTURES CONSIDERED AS MEANS OF EXPRESSION.

Definition of a posture—Simplicity of study—Historical records of postures—Postures of all parts—A change of posture is movement — A posture is due to resultant action of muscles and their nerve-centres—It is a direct mode of expression—Free or disengaged parts most expressive—A limb labouring is not susceptible to mental expression—Organic postures, as from difficult breathing—Postures due to gravity—Effect of gravity on plants—Gravity acts differently during sleep—It can affect the postures of the face—Postures due to reflex action—Spontaneous postures—Fallacies—Classification and analysis of postures—Coincident postures—Symmetry—Postures in art—Postures in animals—Postures in plants—Summary.

THE term "posture" indicates the relative position of the several members of the body with regard to one another and the body in general, or the relative position of the individual parts of a member. The study of postures as means of expression is, in some respects, simpler than the study of movements, as means of expression; for a posture is a condition of quiescence, and, as such, is more easily observed, described, and analyzed. Further postures can be represented by verbal description, by casts, photographs, or drawings; they have been represented in works of art from ancient times, in statues and in drawings, in wall decorations, and on pottery, so that the history of expression by postures can be studied with considerable completeness and precision. The posture of a limb depends upon the relative position of the bones of the limb; and this depends in its immediate mechanism upon the result and action of the opposing muscles which move the limb, the relative tone of the antagonistic flexors and extensors, the adductors and abductors, the pronators and supinators, etc. The posture of the limb is the result of the balance of the opposing muscles. It may be said, then, that a posture is the resultant action of the balance of the opposing muscles which move a part of the body. Taking this meaning of the term posture, we may speak of postures of other parts of the body besides the limbs. The eyes are members moved by opposing muscles; we may, then, speak of their relative position to one another, and the position of each with regard to the axis of the orbit, as the postures of the eyes. In the face the different parts are called features. The cheeks, the openings of the eyes, nose, and mouth, are features of the face.

The features of the face are moved by opposing muscles, so that it is convenient to speak of the postures of the face, or facial postures, as the result of the action or tone of the facial muscles. We may also speak of postures of the head and trunk.

A change of posture is the effect of movement; the posture is the result of the last movement; the cause of the last movement is therefore the cause of the posture. We may describe in anatomical terms the open extended hand and the closed hand; to say that the hand changes from one posture to the other is one method of describing the movement that has occurred. If movements are expressive, postures must be expressive for the same reasons.

Postures, like movements, may be taken as indices of the action, and of the condition of the central nerve-system, because they are the result of the action of the central nerve-system upon the muscles. The posture of the hand, or other part, is the result of the balance of the muscles of the part. The position of the bones of the limb is the resultant of the action of the muscles, and is regulated, to a large extent, by the nerve-system. To consider one part only, let us take the hand and forearm. I have hitherto spoken as though only one stimulus could come from the nerve-centres to the muscles of the limb at the same time; it is probable that many nerve-centres, or portions of the central mechanism, are together sending stimulating currents to the muscles, and that the balance of muscular action is the result of the balance of the action of many nerve-centres. It is, however, not necessary for our purpose to enter upon this discussion as to whether one or many nerve-centres cause a posture; we can proceed, neglecting this point and making the admissions that the balance of the muscular action which regulates the posture of a limb is the outcome of some portion of the central nerve-mechanism. A posture, when thus produced, is a direct expression of the action of some part of the nerve-mechanism.

We will suppose that the right arm, as in Diana (see Fig. 34), is holding a spear firmly; a strong nerve-current is being sent to the muscles of the right forearm, causing the hand to grasp the spear. In order that this position of the hand may change, the stimuli coming to the muscles of the limb must change; they may be all reduced in strength, or their relative strength may be altered. If the woman is startled, she may drop her spear; if she meet a friend, she may purposely throw it away, and then the hand is free to be acted upon again by its nerve-mechanism, the action in the hand produced by the central organ being thus directly expressed. It is in a free and disengaged limb or part that we best see direct expression. The hand of an energetic speaker, if it is not engaged in leaning on the table, or holding a paper, etc., will often express by its movements the general mental state of the speaker. The limb is moved by the strongest motor stimuli that come to its muscles. We see, then, that in the case supposed, the strongest motor impulses coming to the muscles of the limb when it is disengaged, are in some way connected with the mental condition of the subject. If, on commencing to speak, the man hold his hat in his left hand, it is probable that he will continue to hold it throughout his speech. The motor currents to the left side are continuing uninterruptedly, but it may be that (as it is said) his emotion may be so strong, that the current produced by emotion proceeding to his arm will be stronger than the original current; then he will drop his hat and gesticulate with his arm. It is, then, in free and disengaged limbs and parts that we see the best examples of direct expression of mental conditions. Examples of free and disengaged parts may now be compared with examples of the engaged stimulated condition of the same parts.

The hand of a labourer is seen engaged in digging with his spade; his nerve-muscular energy is expended in holding and driving his spade. It would, under such circumstances, require a very strong nerve-current sent to those muscles to alter this forcible stimulus. Hence, the hand engaged in digging is not very impressionable and expressive of the finer motor actions of the nerve-mechanism. When the man puts aside his spade, and talks, especially if at rest, his hand gesticulates and expresses his emotions.

The head is a part or member of the body usually disengaged, and easily moved by slight stimuli; hence movements and postures of the head are usually highly expressive. The head may, however, be not free to move under a slight stimulus, as when a costermonger or fisher-woman carries a basket on the head.

The eyes move freely in their orbits, and their movements give much expression. The eyes may, however, be strongly attracted to an object, as a light (this is probably a reflex action). When the eyes are thus fixed, they are not easily moved by slight stimuli, and cannot be said to be free and disengaged.

The face, again, is a region in which the muscles

Fig. 17.—Dying Gladiator.

are usually free to be stimulated to contraction by slight causes. The capacity of the face for expression exceeds that of any other part of the body. The mobility of the face varies greatly, and there are circumstances under which the facial muscles are not free for the expressions of emotion, as when eating, and when a strong light causes spasmodic contraction of the orbicular muscles of the eyes.

Postures of the body may be due to organic conditions; they may be due to conditions of the organs, and not express the simple action of the nerve-mechanism. Thus, in spasmodic asthma, the patient, if lying in bed at the commencement of the seizure, is obliged at once to rise up, and he sits leaning forward, with his knees drawn up, his elbows on his knees, and his head supported by his hands, labouring for his breath. Various postures are assumed to facilitate respiration; the patient stands erect, with his head thrown backwards, seizing some object to give greater vigour to his efforts, or he leans the head forward on his hands. Such postures are assumed to enable the respiratory muscles to act with greater mechanical advantage. In heart-disease the patient is often unable to lie down, and sits in bed, or in his chair, supported by pillows. Sir Charles Bell* draws attention to postures resulting from organic conditions in his criticism upon the Dying Gladiator (see p. 303).

Gravity is often a factor of great potency in producing postures of the body. All living bodies are under the influence of the same laws of gravity

* *Op. cit.*, p. 194.

as non-living things. The general effects of gravity as affecting the growth and development of animals have been referred to by many authors, but the effect of gravity as a stimulus upon growing or mobile parts in animals has been but little investigated. In plants gravity influences growth in such a manner as to produce various movements.*

Gravity may be a factor in causing postures in the human body in two different ways.

Gravity produces a tendency in a limb to fall downwards, *i.e.* to place itself with its centre of gravity as low as possible. Thus, when the muscles of a man's arms are not stimulated to action by nerve-currents, and still the man stands erect, the arms, under the influence of gravity, fall by his sides. When in a strong man the head bends to one side, it tends to become erect again; that is, the muscles tend to resume such a balance that they are in equal tension on the two sides. This may receive some explanation on the supposition that when gravity causes the muscles on the convex side to be strained, a nerve-stimulus is received by those muscles through reflex action causing them to contract. It must, however, be borne in mind that the facts concerning growth of seedling plants show that living cells may be directly stimulated to certain kinds of action by gravity.

The face is affected by the action of gravity when paralyzed or passive. The jaw drops when the masticatory muscles are relaxed, and when the facial muscles are paralyzed, the tissues of the face

* Prantl and Vines. "Text-book of Botany," p. 87.

fall in their relative position to the skull which supports them. This is easily observed in cases of palsy of one side of the face, where the position of the features on the two sides is easily compared.* Postures of the face may then in part be due to gravity. As far as I know, the positions of the eyes are not affected by gravity. One of the indications or expressions of a moderate conditions of sleepiness or debility is that the head is not kept erect; gravity fails to stimulate the muscles put on a slight strain by bending (inclining) the head to one side.

In looking to the significance, meaning, or direct expression of postures, we must try and determine the causation. Some postures are the result of reflex action.

If a small object is placed in the hand of a healthy child, a year old, the fingers close on it. This is probably a purely reflex movement. The posture of the hand or fingers which results, being due to reflex action, may be termed a reflex posture, in contradistinction to the so-called spontaneous postures. Examples of reflex postures are seen in contraction of the orbicularis oculi under the influence of light; in many cases where an object is grasped in the hand without a conscious act on the part of the subject; in the arm when kept near a part of the trunk that is the seat of some irritation (being tickled).

Spontaneous postures are those which appear to come about as the result of the intrinsic action of

* See chap. xi.

the nerve-centres, and which we cannot at present classify as either reflex or voluntary. The postures assumed in an infant, who probably at birth has no volition, are spontaneous if they are not reflex. Postures in the adult are termed spontaneous if they are not known to be either reflex or voluntary, *i.e.* accompanied by consciousness. Postures assumed by the subject when unconscious, as from the influence of chloroform, may be said to be spontaneous, the outcome of the action of the nerve-mechanism.

The postures that we see assumed under various emotions, when we have evidence that they are not purposely assumed, are indications of the action of the nerve-centres; hence the value of the study of postures in children, whose movements are often so little the result of self-consciousness, and are so commonly purely spontaneous.

In certain other cases postures are not indications of the condition of the nerve-system. Postures of the hand are often due to chronic changes in the joints, the ends of the bones which are in apposition becoming changed as the result of disease, and the ligaments so thickened and contracted as to interfere greatly with the action of the muscles upon them. If the muscles cannot freely move the joints, the stimulus from the nerve-centres cannot be accurately expressed by the postures of the limb. In many cases of crippling rheumatism the postures are caused by the local joint condition and indicate the state of the joints, not the condition of the nerve-centres. So also diseased conditions of the muscles may prevent expression by nerve-muscular signs.

Another set of cases must be mentioned. The postures observed may be the result of a defect of the nerve of the limb, or of the part of the cord or brain with which those nerves are connected. In such cases the postures assumed are expressions of the chronic diseased condition of the nerves or their central origin, and mobile expressions are not indicated.

Voluntary postures imply voluntary movements. We all have some idea of the difference between voluntary and spontaneous movements. It may be said that a voluntary movement is an act of volition, and is preceded by a condition of consciousness. We need not discuss the question here; it will be referred to again in the chapter on " Physical Signs of Mind." We all know that postures can be voluntarily assumed by an exercise of the will. Voluntary postures are objective signs indicating an effect of the will. It is not proposed that we stop to inquire here what the will is—whether it be a function of the brain or otherwise; we are here only concerned with its expression in objective nerve-muscular signs, the motor manifestation of the physical basis of the will.

Grasping a spear, holding the hand of a friend, are nerve-muscular physical signs, and doubtless they express feelings and volition in the subject. We are here concerned to describe and analyze the outward signs of expression—the posture of the hand holding the spear, or grasping the hand of the friend. It appears, then, to be impossible to define exactly what a voluntary posture is; we cannot give definite criteria by which we may distinguish a voluntary posture from one that is reflex, or the spontaneous outcome of the action of the nerve-centres. We fail, then, to define the adjective "voluntary" as applied to a movement, and this failure is due to our want of knowledge of the criteria of volition. So we shall fail again and again when we try to determine logically the physical criteria of mind. We may accept certain criteria as evidence of mind; we cannot prove those signs to be evidence of mind.

Something ought to be said now about the classification and analysis of postures. Now, as postures are nothing but the results of movement, the principles which enable us to analyze and classify movements ought to enable us to analyze and classify postures. It appears needless to repeat those principles verbally, but they may be shown to be applicable to analysis of postures by means of the table in chap. ix., p. 172.

Coincident postures are often seen in practice, thus: when the left eye is directed outwards, the right eye is directed inwards, and *vice versâ*. Symmetry of posture is a division of coincident postures. Certain postures of the head and the hand so commonly coincide as to give special expression. In a weakly child we often see the left hand in the nervous posture, and the head in slight right rotation with right inclination and flexion. In that physical condition where the mental state is called "horror" at the sight of an object, the upper extremities extend towards the object, the eyes and head rotate from it.* To speak of the coincidence of two or more postures is to speak of a member in the series of postures. We speak of coincidences of posture just as we speak of coincidences of movement.

The study of hemiplegia and hemispasm, in comparison with examples of coincident postures, and movements which are highly expressive, suggested to me to look for facts concerning coincident postures such as are recorded above.

In art, movement is expressed by the postures of the subject. The outstretched hand expresses energy in movement, for the limb could not so remain unless force were expended. Emotion is often expressed by posture. The head is hung in shame, extended in adoration, rotated with flexion and inclination in nervousness.

Darwin, in his work on "Expression," has shown that probably the snarl of the dog is expressive of similar emotion to the sneer of a man, and he produces many examples of analogy between expression by postures in animals and in man.

Lastly, plants, in their parts which move, have expressive postures. The *Mimosa* leaf expresses recent excitation; the infolded tentacles of the leaf of *Drosera* expresses that an insect or piece of nitrogenous food has been caught by it.

Summary.—To summarize what has been said in this chapter; postures are the results of movements, and for that reason they have the same physiological significance as movements; that is, postures

---

* Compare this with what was said about head movements in hemiplegia (see chap. vii.).

are nerve-muscular signs. A posture is the position of a part of the body or a member, and is due in its mechanism to the balance of its muscles. We may, then, speak of the postures of any members of the body, head, face, eyes, etc. It is in free or disengaged members, or parts, that we see the best examples of direct expression. A hand engaged in labour is less expressive of mental conditions than an unoccupied hand. Postures may be determined by organic conditions rather than by the state of the nerve-centres; thus, the body is supported, when there is urgent difficulty of breathing. In the Dying Gladiator the body is supported by the arm to assist the breathing. Gravity often determines the position of the head and limbs, when there is no strong nerve-current proceeding to its muscles.

The position of a member may be due to reflex action. Observation shows that a special posture follows uniformly upon the stimulation of a particular sensory surface; here the posture is not the spontaneous outcome of what originates in the subject, and the posture is less likely to indicate the volition or the brain condition. Free and disengaged parts are alone expressive of brain conditions.

Postures may be due to organic disease, or to changes in the joints; in such cases they are not direct indices of the brain.

The effects of gravity are worthy of study; they are better understood in plant-life than in man.

Postures may be classified in the same manner as movements. We see examples of postures alike in man, animals, and in plants; in each case such signs are expressive.

CHAPTER IX.

POSTURES OF THE UPPER EXTREMITY.

Method of examination—Anatomy—The convulsive hand contrasted with the hand in fright—The feeble hand and the hand in rest—The straight extended hand, normal—Application of the principles of analysis—Straight extended hand with the thumb drooped—The nervous hand, seen in art—Energetic hand the antithesis of the nervous hand—Table giving analysis of postures, and application of the principles—Principles of analysis—Anatomical analysis—Small parts contrasted with large parts—Interdifferentiation—Collateral differentiation — Symmetry — Excitation of weak centres—General excitement or weakness — Analogy—Antithesis—Fallacies—Methods of determining whether a posture is the outcome of the spontaneous action of the nerve-centres.

I PROPOSE here to speak of postures of the arm, forearm, and hand. The postures that have been frequently observed in actual life will be described, and their significance will be discussed afterwards. I have made observations upon these points in some thousands of subjects—in healthy subjects, and among my patients, males, females, and children. I have, as the result of these observations, been led to accept certain postures of the upper extremity as expressions of certain conditions of the individual, as physical signs, or objective observable expressions of the conditions inherent at the time.

In making such examinations while the subject is standing, he is requested to hold out his hands, or sometimes to hold out the hands with the palms downwards. This affords the opportunity of observing the free or disengaged hands. The postures, the symmetry, presence of any movements, the coincident balance of the head or spine, etc., can be noticed while observing specially the hands.

The upper extremity in man consists of the following parts:—The upper arm, with the humerus as its bone, which is articulated at the shoulder with the scapula or blade-bone. The lower end of the humerus articulates with the radius and ulna, the bones of the forearm. The forearm articulates with the wrist, which is made up of eight small bones, collectively spoken of as the carpus. The palm of the hand consists of five metacarpal bones, each of which carries its own digit. These bones, termed collectively the metacarpus, are capable of slight movement, enabling them to be approximated like a bundle of sticks tied together, or spread out. The junction of a metacarpal bone with its digit is called the metacarpo-phalangeal joint, or the knuckle. The individual bones of the fingers and thumb are termed "internodes."

For the general convenience of description, I have illustrated and tabulated eight typical postures indicative of positions commonly seen.

Perhaps one of the best known spontaneous

postures of the hand due to the brain condition, is the "convulsive hand."

This was well described by Trousseau[*] as a common condition of the hand in tetany. "In the upper limbs, the thumb is forcibly and violently adducted; the fingers are pressed closely together, and semiflexed over the thumb in consequence of the flexion of the metacarpo-phalangeal articulation; and the palm of the hand being made

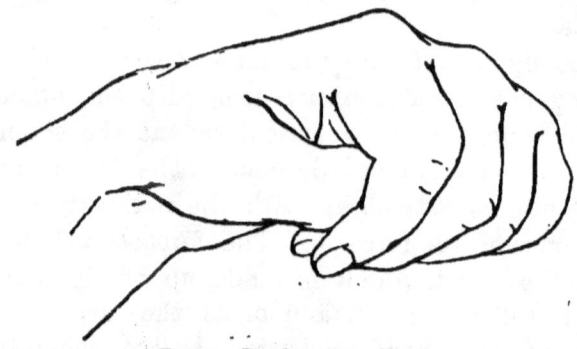

Fig. 18.—The Convulsive Hand.

hollow by the approximation of its outer and inner margins, the hand assumes a conical shape."

The condition of the metacarpus is specially noteworthy. I have taken casts of several cases and have measured many more. It is, as Trousseau described it, arched or contracted by approximation of the metacarpal bones, which are screwed together. Important variations of this posture might be described; it is often seen in various pathological states.

[*] "Clinical Medicine," vol. i., New Sydenham Society's translation, 1867, p. 375.

ANALYSIS OF THE CONVULSIVE HAND. 157

Let us now analyze this posture due to tetany, and observe and describe the characteristic constituents of the posture, and see if these elements occur in other less-marked cases.

The small parts, the digits, and the larger parts are all in flexion; there is a general struggle among the muscles, and the stronger move the bones; the nerve-centres appear to be all in activity, and the

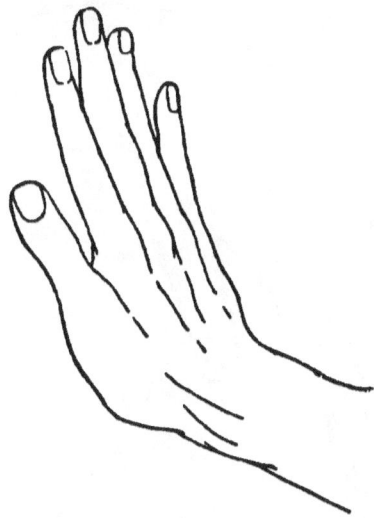

Fig. 19.—The Hand in Fright.

stronger prevail in causing the general flexion. This posture is antithetical to the hand in fright.

	Hand in fright.	Hand in convulsions.
Wrist	Extended	Flexed
Metacarpo-phalangeals	Straight extended	,,
First and second internodes	,, ,,	,,
Thumb, metacarpo-phalangeal	,, ,,	,,
,, internode	,, ,,	,,
Phalanges, relative position	All in same plane	Contracted and adducted.

The general strength of the nerve-discharge is indicated by all the strong centres being stimulated, whereas in the antithetical posture the weak extensor centres are stimulated.*

Let me say here, that in using the terms "convulsive hand," "hand in fright," etc., I do so only as convenient terms, and do not wish to make the tacit assumption that the hands are always in this

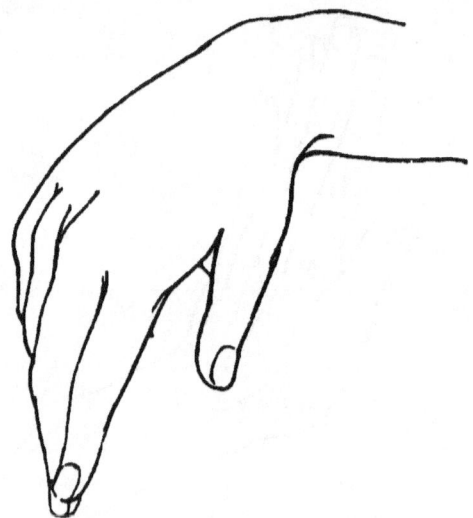

Fig. 20.—The Feeble Hand.

posture in the condition of convulsion or in the emotion "fear."

Now to discuss the posture of the "hand in fright" (Fig. 19). The anatomy of the posture has been described. The small parts, the phalanges, as well as the large joint, the wrist, are alike extended; the joints further from the trunk, as well as those

* The contracting or drawing together of the metacarpal bones is an element in this posture, seen also in the "feeble hand."

nearer, are in the same state; the collateral parts, the fingers and metacarpals, are all in similar relation; there is no collateral differentiation.

The feeble hand.—If the disengaged hand of a feeble child or woman is held out, we commonly see a slight, but important, departure from the hand in rest. The thumb is drooped, and its metacarpal bone is approximated to the palm, all the metacarpals being bent round as in the convulsive hand. This is a posture often seen.

The hand in rest may be seen in a strong and

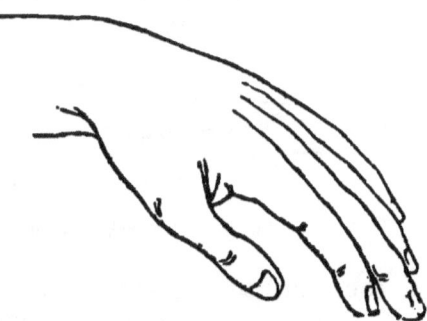

Fig. 21.—The Hand in Rest.

healthy subject during sleep, also when the hand hangs free and quiet, as when doing nothing, and in a state of mental quiescence. I have often seen it during quiet conversation, and in men while walking or travelling. The metacarpus is often slightly contracted, but not to such a marked degree as in the "feeble hand."

The "feeble hand" and the "hand in rest" have the same anatomical description, but they differ in degree. It was the result of clinical observation that induced me to describe the two forms. The hand in rest is normal in the condition of resting; the feeble hand is hardly to be considered as normal, at any rate in a man. The principal difference in the two postures is in the metacarpus; in the feeble hand the contraction of the metacarpus is marked, and this I believe to be abnormal, because I do not see it in a strong subject, and because it is often seen in convulsion and in chronic rigidity sequent to paralysis, etc.

The normal straight hand when held out is the posture typical of strength. This is the posture in which a strong and healthy man, woman, or child

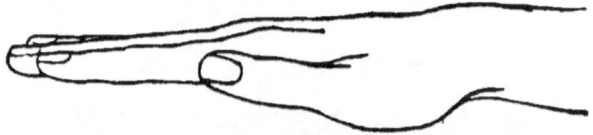

Fig. 22.—The Straight Hand.

holds out the hand to request; there is no flexion, no extension, but perfect balance. This statement is founded on numerous observations. I have often requested a body of students, or other healthy subjects, to hold out their hands; so also healthy children, who could know nothing of my object; and thus I have had the opportunity of judging that this well-balanced posture indicates the normal action of the nerve-mechanism employed in the act.

There is no interdifferentiation, no collateral differentiation, no difference of small parts in contrast with large parts; the metacarpus is straight transversely, in contrast with the contracted metacarpus of the feeble hand and the hand in rest.

Now, the first degree of deviation from this ortho-extended hand, which is the type of strength, is seen in the posture of the thumb metacarpal bone. The "ortho-extended hand with thumb drooped"—

Fig. 23.—The Straight Extended Hand with Thumb drooped.

this is a departure from the normal in the direction of weakness, the first and most mobile, and the most specialized bone of the metacarpus being adducted, as in the convulsive hand, the feeble hand, and the hand in rest. This is often seen in a man who is strong and well, but tired with a day's work.

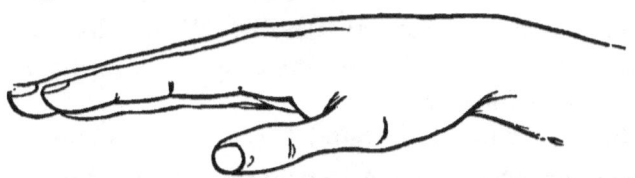

Fig. 24.—Hand intermediate between the Hand in Rest and the Straight Hand.

The two next modes of expression by hand postures that have to be described are those which earliest attracted my especial attention. It was the clinical value of a knowledge of the "nervous hand" that encouraged me to pursue these investigations on expression.

Having during some years given special study to the conditions of the nerve-system in children, my attention was especially drawn to the various postures presented by children brought to me for examination at the East London Hospital for Children, and from 1878 I kept notes of the spontaneous postures observed.* The children were requested to hold out their hands, and the passive condition or posture of each hand was noted. At first it was difficult to describe the postures in anatomical language, though some were seen to be characteristic of certain nerve conditions. In 1879, while visiting Florence, it struck me that the posture of the hands of the Venus de' Medici † was exactly similar to the posture so often seen in nervous children. Later in the year, at the British Museum, I saw the English Venus side by side with the Diana—feminine coyness and nervousness represented side by side with the expression of energy and strength—and the contrast of the hand postures showed them to be in direct antithesis. While looking at the marble hands it became easy to describe their anatomical postures.

In the "nervous hand" the wrist is slightly flexed or bent, the metacarpo-phalangeal joints are moderately hyper-extended (extended beyond the straight line), the first and second internodes being either slightly flexed or kept straight. The thumb is extended backwards, and somewhat abducted from the fingers. This spontaneous posture I have

---

* "Spontaneous Postures of the Hand considered as indications of the Conditions of the Brain." Read before the Royal Medical and Chirurgical Society of London, November 28, 1882. See "Brain," part xxiii.

† See Fig. 32, p. 296.

seen, and others with me, in hundreds of cases usually in nervous children, bad sleepers, those convalescent from chorea, etc. The same posture is sometimes seen in partial hemiplegia. The posture is often bilateral, but is usually unequally represented on the two sides; it is also often seen on one side only, especially in children convalescent from hemichorea.

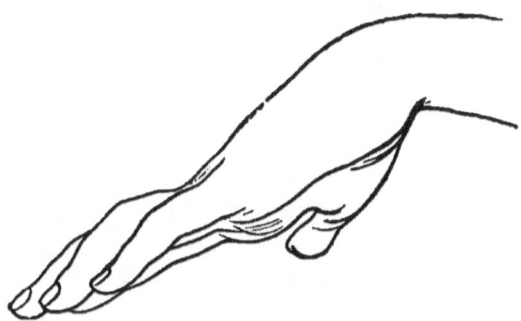

Fig. 25.—The Nervous Hand.

Another posture of the hand less frequently seen in pathological states, I described under the name of the "energetic hand," and since so doing I have found the posture figured by H. Meillet of Paris, with the note "Main dite du prédicateur emphatique. La Salpêtrière, service de M. Charcot, salle Saint-Paul, No. 6, Isméric Angot." It is there figured as a permanent deformity resulting from brain disease. Dr. Little also figures this posture as due to spastic contraction. In this posture the wrist is extended, and the small joints are all in flexion. Here analysis shows the large joint, the wrist, in extension, the opposite to the condition of weakness, and this extension of the large joint gives to the posture the indication of excitement in the nerve-mechanism. The small joints are all flexed, as seen in the hand of a man in sleep or resting, and this gives the posture the indication of strength or activity with rest.

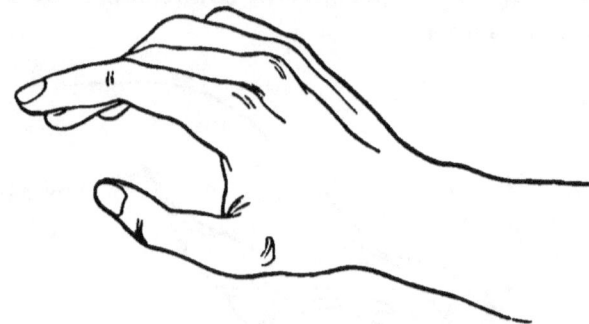

Fig. 26.—The Energetic Hand.

Anatomical analysis shows the "energetic hand" to be the antithesis of the "nervous hand."

	Nervous Hand.	Energetic Hand.
Wrist	Flexed	Extended
Metacarpo-phalangeals	Extended	Flexed
First and second internodes	Flexed	,,
Thumb, metacarpo-phalangeal	Extended	,,
,, internode	,,	,,
Phalanges, relative position	Slightly abducted	Adducted

The physiological expressions of the two postures are also opposites.

This suggests another principle of use in explaining the meaning of certain postures. The principle of antithesis * may be enunciated as follows: "In

---

* Charles Darwin uses the principle of antithesis to explain many modes of expression. See "Expression of the Emotions," p. 28.

opposite conditions of the nerve-mechanism producing the postures of a joint, the postures will be opposite or antithetical;" *e.g.* in a weak or inactive condition of the nerve-mechanism governing the wrist, flexion results; this would lead us to anticipate that in an excited condition of this piece of nerve-mechanism we should find the wrist extended.

The postures of a limb depend in their immediate mechanism upon the resultant action of opposing muscles, the relative tone of the antagonistic flexors and extensors, the adductors and abductors, etc. Various views may be held with regard to the nerve-mechanism which regulates the contraction of opposing sets of muscles. That mechanism may be considered a line of reflex action regulating the action of the muscles, or the balance may be considered due to some other kind of nerve-mechanism. In any case it will, I think, be granted that some portion of the central nerve-system is the cause of that balance of the muscles that produces the posture, and therefore the posture is an index of the condition of that central nerve-mechanism.

In examining the condition of the nerve-system in children, it has for several years been my habit to observe what spontaneous posture would be assumed by the hand when the forearm was held out.[*] Let a weak, nervous child be requested to hold out her hands in front on a level with the shoulder in a prone position. The limb is now

[*] See "Brain," part xi.; and *British Medical Journal*, December 6, 1879.

free or disengaged, and the posture assumed by the hand is in most cases, I believe, the spontaneous outcome of the action of the nerve-centres.

From continued observation of spontaneous postures, and after frequently trying what spontaneous posture would be assumed by the free hand in various subjects, cases of weakness, strong men, feeble women, nervous children, an empirical idea was obtained as to the indication of each posture. Then comparison, analysis, and analogy enabled me to suggest some definite principles. In seeking a rational explanation of this posture, I first tried to find a solution by looking at it after the manner in which Charles Darwin made most of his observations upon postures. Cases were looked for where this posture of the hand was assumed or brought about for some useful purpose, or its occurrence was attended with some kind of amelioration of weakness. Then this "nervous hand" was seen, in a hand not free or disengaged, in people who, standing, lean a little forward with the outspread hands resting on the table for support. This hand seeking rest droops at the wrist, and then if the stimulus to the muscles is weak, as the weight of the trunk bears upon the resting hands, hyper-extension of the metacarpo-phalangeal joints and of the thumb is mechanically brought about. If this be accepted as an explanation, the argument stands thus :—

1. Observation shows that when a man is tired or weak he often rests his hands as described.

2. The nerve-mechanism corresponding to the hand is thrown into the special condition corre-

sponding to the "nervous hand," and thus that posture is mechanically brough about.

3. The nerve-mechanism corresponding to the hand falls spontaneously into the condition corresponding to the "nervous hand" when the man is tired or weak.

4. When the hand of a weak man is held out free, its nerve-mechanism spontaneously places it in the "nervous posture."

The attempt at explanation may be made again, commencing with the principle of analysis and then applying the principle of analogy.

Anatomical analysis shows that this "nervous hand" presents two principle elementary conditions:

(1) Flexion or drooping of the wrist.

(2) Hyper-extension of the metacarpo-phalangeal (small) joints.

So much for analysis, now as to analogy.

Looking for analogy between postures, flexion or drooping of the wrist may be seen, as in a hand that is resting, in a hand that is passively held out while the patient is deeply asleep, or in deep coma, or in a paralyzed arm. It is probable, then, that the wrist flexion indicates the weakness of the "nervous hand."

Hyper-extension of the metacarpo-phalangeal joints and thumb may be seen as a temporary condition in subsultus tendinum and in chorea. In all these cases the unstable condition of the nerve-mechanism is indicated by the extensor movements of small parts.

Now, looking at the elementary conditions shown

by analysis to constitute the "nervous hand," and the probable indication of the conditions analogous to these, we get some kind of explanation of this posture. The drooping wrist is analogous to that seen in hemiplegia; the hyper-extension of the knuckles and thumb is analogous to conditions such as chorea and subsultus tendinum. This appears a more rational explanation than that first given, and the methods of analysis and analogy give some indication of the physiological significance of the posture. Something very like this posture of the hand is indicated by Dr. Little in his work on deformities, 1853, as the result of "spastic contraction."

My present object is to show that spontaneous postures may be studied with advantage as indications of the conditions of the brain. As to the principles involved in these inquiries. A primary postulate, involved in the following principles, is that—

"If we see some spontaneous nerve-muscular action often repeated in the same and in different subjects, it may be assumed that there is some nerve-centre, or nerve-mechanism, corresponding which can act independently." In any particular instance the assumption of the existence of such a centre or piece of nerve-mechanism would be strengthened if it could be shown that there are cases of excitation and cases of paralysis of such hypothetical centre.

The following suggestions are offered as to the *methods of determining whether a posture is the outcome of the spontaneous action of the nerve-centres.*

A posture observed can only be considered as the outcome of the spontaneous action of the nerve-mechanism when the limb or part is free and disengaged. If the muscles of the hand are engaged in holding an object, or in an act of manipulation, the postures of the hand are not simply the outcome of the spontaneous action of the nerve-mechanism.

Postures are frequently seen accompanying, and apparently caused by, demonstrable brain-disease; as examples, the "convulsive hand," cerebral facial palsy.* When these postures occur otherwise, apparently spontaneously, they may be looked upon as nerve-muscular signs produced by the nerve-mechanism corresponding.

Postures which are frequently seen in very young children and infants are certainly spontaneous. Postures seen to be frequently repeated in many such young subjects are probably spontaneous nerve-muscular actions. With regard to postures thus observed in many infants, it is probable that the piece of nerve-mechanism corresponding is well developed, and rendered very definite so as to be easily excited to spontaneous action.

When a posture seen in an adult is found to be analogous to one often seen in infants, the probability of its spontaneous origin is strengthened.

Postures of disengaged parts, still observed in man, and frequently represented in ancient art,†

---

* See pp. 108, 109, Figs. 10, 11.
† See Fig. 35, p. 300.

are probably of frequent spontaneous occurrence, and highly characteristic nerve-muscular postures.

The application of the "principles of analysis" may assist in determining whether a certain posture observed is probably the outcome of the spontaneous action of the nerve-centres.

In these studies care must be taken to avoid certain errors from attributing all postures to the action of the central nerve-mechanism.

Postures may be due to joint-disease, especially arthritis deformans. M. Charcot carefully differentiates between this deformity and the posture he has described as the "writing hand" often seen in paralysis agitans.

Postures may be due to local causes—tumours, inflammation, etc.

Postures may be determined by organic conditions, such as difficulties in the respiratory and circulatory organs causing orthopnœa. Sir C. Bell drew attention to this matter in his critical analysis of the posture of the Dying Gladiator.* Such postures are not the direct effect of the spontaneous action of the nerve-centres.

Gravity may be a factor in the causation of postures, whether the member be "free" or "engaged," thus: in the figure of Hercules at rest,† he leans on a vertical club to support his body, and the posture of the right arm is determined mainly by gravity. In a paretic arm such as is often seen in chorea, or from brain disease, the wrist will droop into flexion if the forearm is held out

* See p. 303. † See p. 305.

prone, but will fall into extension if the forearm is held out supine. Dorsal decubitus results from gravity, together with general palsy of the motor nerve-mechanism.

Postures may be due to local nerve injury or disease, thus: injury to the facial nerve;* injury to the musculo-spiral nerve; a gumma pressing upon the third cranial nerve, may produce certain postures of the parts supplied by these nerves, such postures not depending upon the condition of the nerve-centres.

Postures may be due to rigid muscular contraction dependent upon permanent brain lesion, such as descending sclerosis. Such cases should be kept separate from the postures due to a temporary, it may be momentary, condition of the central nerve-mechanism. It is hoped that some proofs have been given that the study of spontaneous postures as indications of the condition of the brain is useful, and that it may be considered as one of the exact methods of studying the nervous system in its physiological and pathological conditions.

The subjoined tables contain descriptions of eight typical postures, applying the "principles" to the study of each.

* See Fig. 27, p. 202, representing facial palsy.

TABLE 1.

Principles.	Nervous Hand.	Energetic Hand.	Hand in Rest.	Ortho-extended Hand.
1. "Anatomical analysis."	Wrist . flexed. Metacarpo-phalangeals hyper-extended. 1st and 2nd internodes flexed. Thumb, metacarpo-phalangeal hyper-extended. Thumb, internode hyper-extended. Phalanges, relative position abduction.	Wrist . . extended. Metacarpo-phalangeals flexed. 1st and 2nd internodes flexed. Thumb, metacarpo-phalangeal . flexed. Thumb, internode flexed. Phalanges, relative position . . adducted.	Wrist . . . flexed. Metacarpo-phalangeals flexed. 1st and 2nd internodes flexed. Thumb, metacarpo-phalangeal . flexed. Thumb, internode flexed. Phalanges, relative position . adduction.	Wrist. ortho-extended. Metacarpo-phalangeals ortho-extended. 1st and 2nd internodes ortho-extended. Thumb, metacarpo-phalangeals ortho-extended. Thumb, internode ortho-extended. Phalanges, relative position adducted.
2. "Small parts contrasted with large parts."	Small parts characterized by hyper-extension of the metacarpo-phalangeal joints in contrast with flexion of the large joint—the wrist.	Small parts in flexion in contrast with the flexion of the large joint—the wrist.	Small parts in flexion. Large parts in flexion.	Small parts ortho-extended. Large parts ortho-extended.
3. "Interdifferentiation."	Interdifferentiation marked, the small metacarpals and the thumb being extended in contrast with flexion of the large joint—the wrist.	Interdifferentiation marked, the small joints being all flexed in fingers and thumb in contrast with extension of the large joint—the wrist.	No interdifferentiation in the type described.	No interdifferentiation in the type described.

TABLE I. (continued).

Principles.	Nervous Hand.	Energetic Hand.	Hand in Rest.	Ortho-extended Hand.
4. "Collateral differentiation."	No collateral differentiation in the type described.	No collateral differentiation.	No collateral differentiation.	No collateral differentiation.
5. "Symmetry."	The type symmetrical; the two hands may differ.	Symmetrical.	Symmetrical.	Symmetry of this type complete.
6. "Excitation of weak centres."	Metacarpo-phalangeals extended, indicating excitement of weak centres.	Wrist extended, indicating excitement of the weaker centre of a large joint.	No extension of any joint, indicating no excitement of weak extensor centres.	Perfect balance, no excitation or weakness. This is the normal strong hand.
7. "General excitement or weakness."	Cannot apply here, as there is interdifferentiation.	Cannot apply here, as there is interdifferentiation.	General flexion indicating predominance of the strong flexor muscles from general weakness of the nerve-centres.	Perfect balance, no excitation, no weakness.
8. "Analogy."	Wrist flexed as in "hand in rest." Knuckles and thumb in hyper-extension as seen in chorea, in subsultus tendinum in some irregular twitches in sleep.	Wrist extended as in "hand in fright." Knuckles and fingers and thumb flexed as in "hand in rest."	The type of complete relaxation of all the nerve-centres.	The type of complete balance of all the nerve-centres, hence the type of strength without excitement.
9. "Antithesis."	The energetic hand.	The "nervous hand."	a. The "ortho-extended hand." β. The "hand in fright."	a. The "feeble hand." β. The "hand in rest."

TABLE II.

Principles.	Convulsive Hand.	Hand in Fright.	Feeble Hand.	Ortho-extended Hand with Thumb drooped.
1. "Anatomical analysis."	Wrist flexed or extended, etc. Metacarpo-phalangeals 1st and 2nd internodes flexed. Thumb, metacarpo-phalangeal flexed strongly. Thumb, internode flexed strongly. Phalanges adducted.	Wrist hyper-extended. Metacarpo-phalangeals ortho-extended. 1st and 2nd internodes ortho-extended. Thumb, metacarpo-phalangeal ortho-extended. Thumb, internode ortho-extended. Phalanges abducted.	Wrist . . . flexed. Metacarpo-phalangeals flexed. 1st and 2nd internodes flexed. Thumb, metacarpo-phalangeal . . flexed. Thumb, internode flexed. Phalanges adducted, metacarpus contracted.	Wrist ortho-extended. Metacarpo-phalangeals ortho-extended. 1st and 2nd internodes ortho-extended. Thumb, metacarpal bone adducted or drooped. Thumb, internode ortho-extended. Metacarpus straight, phalanges adducted.
2. "Small parts contrasted with large parts."	Condition of the wrist varies (it may be flexed, extended, adducted).	Small parts ortho-extended, while the large wrist-joint is hyper-extended.	Small parts in flexion. Large parts in flexion.	
3. "Interdifferentiation."	Interdifferentiation various; the wrist posture varies.	No interdifferentiation, all joints extended.	No interdifferentiation; all joints flexed.	
4. "Collateral differentiation."	Probably no collateral differentiation in most cases.	No collateral differentiation.	No collateral differentiation in the type described.	Collateral differentiation of the metacarpal bones, the thumb being drooped.
5. "Symmetry."	...	...	Symmetrical.	Not symmetrical; the metacarpus being straight but the thumb drooped.

POSTURES OF THE UPPER EXTREMITY. 175

TABLE II. (*continued*).

Principles.	Convulsive Hand.	Hand in Fright.	Feeble Hand.	Ortho-extended Hand with Thumb drooped.
6. "Excitation of weak centres."	All the hand in strong flexion, but the wrist may be extended, indicating excitement of this weak centre.	All the extensor weak centres are excited.	No excitement of any weak centres.	No excitement of weak extensor centres, but marked weakness of the extensors of the thumb.
7. "General excitement or weakness."	General excitement of all the nerve-centres causes the flexors strongly to predominate.	General excitement of all the extensor centres, indicating general excitement of the weak centres.	General weakness.	Perfect balance of nerve-mechanism for fingers and wrist, weakness of the centres for the thumb.
8. "Analogies."	...	α. Analogous to the "energetic hand," with excitement of the weak extensor centres of the digits. β. Analogous to the "ortho-extended hand," with excitement of the weak extensor centre of the wrist.		
9. "Antithesis."	Metacarpus contracted as in the "feeble hand."	Hand in rest.	Metacarpus contracted as in the "convulsive hand." Ortho-extended hand.	

It seems desirable to give some further indication of the knowledge that it is hoped to gain from the systematic study of postures. I do not attempt to indicate what special portions of the brain are concerned in producing different postures, and do not think such details can at present be given. I cannot localize a motor centre for the "hand in fright;" could we do so, it would afford some evidence as to what portion of brain-mechanism is concerned in the mental condition called fright. In the case of an idiot, whose hands were usually in the convulsive posture, I took casts of the hands, and subsequently had the opportunity of examining the brain, which showed marked defects in the posterior convolutions. It is probable that difference in hand posture on the two sides indicates a different action of the two hemispheres of the brain, and of this I hope to give further evidence on another occasion, founded upon the association of postures. When the hands are held free, a difference in the posture on the two sides, or different movements in the two hands, give evidence that the two hemispheres are in a different condition. These studies show that in many people the nerve-centres are not strictly symmetrical in action, the average postures of the face and hands being asymmetrical, *e.g.* frequent occurrence of one-sided grinning; the nervous hand is often seen on one side only, usually the left; these observations may be conducted in healthy people, and give evidence of the individual's idiosynchronism. The study of postures gives examples of the application of the "principles," enunciated above, and, I think, affords some proof of their practical importance. Most of these points are commonplace clinical considerations, but principles vi. and vii. perhaps require further explanation and defence.

It may be readily admitted that it is desirable and important to know what are the motor indications of excitement or over-action of a nerve-centre, also to know the relative strength of the different motor centres, and what are the outward physical signs of widespread stimulation or weakness among portions of the nerve-mechanism. In numerous cases, where common experience has shown general strength, weakness, or widespread stimulation, etc., these principles of analysis have been applied, and some general knowledge has thus been obtained, but more exact knowledge is wanted. I trust by an experimental method, described in the *Journal of Physiology,** to be able to give hereafter a much more precise and definite account of these principles.

These studies afford evidence that some parts of the brain may be in a condition of weakness, while others are in a state of strength or excitement, and that one hemisphere may be weaker in action than the other. Again, a change of posture may indicate a change in the corresponding nerve-mechanism. Thus in chorea, when the forearm is passively supported, we often see that the free hand passes from the posture of rest to the convulsive posture, thus indicating the change in the condition of the nerve-mechanism, corresponding to the muscles of the

* Vol. iv. No. 2, August, 1883; also see chap. xix.

hand. By daily recording the most common involuntary postures of the free hand, we obtain indications of the daily condition of the nerve-mechanism corresponding.

In analyzing any posture of the hand we may consider the bearing upon the subject of the following principles, or points of analysis. These are the same that have been used in the tables.

I. Every posture can be described in anatomical language, giving the position of each joint of the limb.

II. Consider the posture of the small parts of the limb as distinguished from the larger parts. In chorea the movements of the small parts are much more obvious, and probably much more frequent than those of the larger parts; more movement is seen in the fingers than in the elbow and wrist. A slight amount of brain disease will interfere more with the use of the fingers than with the use of the upper arm. A man may be able to stretch out his hand and grasp an orange, but be quite unable to write, or to pick up a pin from the table.

III. Consider the different relative postures seen in the large and small joints.

Interdifferentiation of the postures of the large and small joints may give important indications; the large joint may be paretic or weak, while the small joints are in extension, from irritability of the nerve-centres which govern these movements. In such a case there would be an interdifferentiation of the nerve-centres governing the different joints. In the nervous hand the wrist is weak, the fingers show irritability.

IV. Consider the relative condition of posture in collateral joints.

The knuckles are collateral joints. The posture of each may be similar, as in "the hand in rest," or it may be dissimilar, as in "the straight hand with the thumb drooped," where the metacarpal bones, which are collateral parts, are four in the same plane, and that of the thumb is adducted. Collateral differentiation is usually seen in the movements of the hand in chorea, also in the spontaneous movements of the fingers of a healthy infant; it is marked in some cases of athetosis.

V. Consider if the posture is symmetrical. Observe whether the postures of the corresponding parts on the two halves of the body be symmetrical, whether the posture of either hand be the same. Much has been said about this in describing postures of the head and face. Perfect symmetry indicates that the two halves of the brain are acting similarly.

VI. Consider if there be any indications of excitement or over-action of any nerve-centre or centres usually weak, such as the centre or nerve-mechanism governing the extensor movement of a joint.

Observations, comparisons, and analysis seem to indicate the nerve-centre, or mechanism governing the extension of a joint (*i.e.* the action of the weaker muscles), as being weaker than the centre governing flexion. Observe if such indications of excitement subside in sleep, or when at rest; they do so in chorea. Some cases, which constantly present the nervous hand when the individual is awake and the limb is free, lose the posture during sleep. In this state, if the forearm be gently held out by an assistant, the hand falls into the posture of rest, but on awaking the patient, the nervous posture is resumed.

VII. If there be general, equal, widespread stimulation or weakness of portions of the nerve-mechanism governing the postures of a limb, or part of the body, the action of the stronger muscles prevails. General weakness may be indicated by general slight flexion, showing all the nerve-mechanism paretic, and the stronger flexor muscles prevailing.

VIII. Consider in the posture if any joint, or set of joints, be in position analogous to that of any other posture, and whether the significance of the position of that joint be the same as in the posture to which the analogy is made.

If we study the analogies of the nervous hand, it seems probable that the wrist flexion has an indication similar to that in the hand in rest. In each case the weakness of the nerve-mechanism governing it is indicated. Now, as to the over-extension of the knuckle-joints, do we see that elsewhere? In chorea, in the finger-twitching of nervous children, in patients exhausted by fever (subsultus tendinum), we often see extensor movements of these joints. This seems to indicate such action as a sign of weakness or irritability, not rest.

Now, looking at the elementary conditions shown by analysis to constitute the "nervous hand," and the probable indication of the conditions analogous to these, we get some kind of explanation of this posture. The wrist drooping is analogous to that seen in rest; the over-extension of the knuckles and thumb is analogous to conditions such as chorea, nervousness, and exhaustion.

IX. Consider if the posture be the antithesis of any other known spontaneous posture.

This principle has been exemplified in contrasting the nervous hand and the energetic hand. It is probable that in many cases opposite conditions of the nerve-mechanism producing postures, will produce antithetical postures. Charles Darwin used the principle of antithesis to explain many modes of expression (see "Expression of the Emotions," p. 28).

This principle of antithesis is also recognized by John Bulwer (see p. 325).

CHAPTER X.

EXPRESSION IN THE HEAD.

Positions and movements of the head defined—Flexion the only symmetrical movement—Action of light in causing head movements; varying effect of such stimulus in different brain conditions—A weak posture—Effect of gravity—The head usually free—Application of the principles of analysis to head postures—Movements of the jaw—Physiognomy, or certain forms of the head—Summary.

IN the study of expression in the head, as in other parts of the body, we must look to the conditions of its development and its special trophic states, and also to its movements and the postures which result from those movements.

The title of Sir Charles Bell's first essay is, "Of the Permanent Form of the Head and Face, in contradistinction to Expression." He goes on to say,* "A face may be beautiful in sleep, and a statue without expression may be highly beautiful; on the other hand, expression may give charm to a face the most ordinary. Hence it appears that our inquiry divides itself into the permanent form of the head and face, and the motion of the features,

* *Op. cit.*, p. 20.

or the expression." And again: "A countenance may be distinguished by being expressive of thought; that is, it may indicate the possession of the intellectual power. It is manly, it is human; and yet not a motion is seen to show what feeling or sentiment prevails."

Here Bell uses the term "expression" as confined to mobile, nerve-muscular, or kinetic modes of expression, and carefully distinguishes between them and the conditions of development which indicate the coincident condition of development of the brain. The permanent forms of the head and face as indicative of the states of the mechanism of the mind have been often described in treatises on physiognomy. Much expression may be seen in the various movements and postures of the head, as well as in its size, form, and proportions. The movements of the head or skull are, no doubt, largely due to movements of the neck, which is made up of the seven cervical vertebræ. The skull can be rotated on the neck; it can be flexed and extended on the first vertebra, which is called the atlas. To simplify description we shall here speak only of movements of the head with regard to the body at large, neglecting the question as to how far the vertebræ of the neck may take part in such movements.

For descriptive purposes we may conveniently speak of two axes of the head: (1) the interparietal, or transverse axis, passing from one ear across to the other; (2) the fronto-occipital or antero-posterior axis, passing from the centre of the forehead to the occiput behind. By referring to the positions and movements of these axes we can define all the positions and movements of the head.

A. *Flexion* and *extension, i.e.* bending forward and backwards of the head, as in nodding. In flexion the antero-posterior axis has its anterior end depressed, but the transverse axis remains horizontal, and the two ears remain at the same level.

In this movement, or posture, symmetry is maintained.

B. *Rotation, i.e.* rotation of the antero-posterior axis in the horizontal plane, the head remaining erect, and the interparietal axis horizontal, without flexion or extension. Such movement may occur in turning the head to look at an object; in this case symmetry is not maintained. Right rotation is used to imply that the forehead moves to the right, as in looking towards the right side.

C. *Inclination, i.e.* depression of one or other extremity of the interparietal axis, in which case the ear on the side of inclination is lower than the other. Right inclination means depression of the right extremity of the interparietal axis, or the right ear; left inclination means depression of the left ear. Inclination is an asymmetrical movement; it may occur without either flexion or rotation, but is commonly associated with both.

The only symmetrical movements of the head are flexion and extension; inclination and rotation are asymmetrical movements involving unlike conditions in the two halves of the brain.

We shall not consider many head movements and

postures, but those that we are about to examine will serve to afford examples of the application of our principles of movements, and in particular they will demonstrate how movements may result from, and be modified by, external agencies. In a strong and healthy man the head is held erect, and symmetrical, unless some central condition or external agent changes the posture. In a strong man the centre of the forehead is in the mid-plane of the body, the antero-posterior and the transverse axes are horizontal, with both ears on the same level: this is a normal posture of quiescence.

Rotation is a movement always involving asymmetry of action; flexion and extension are symmetrical movements.

A slight stimulus, as the sight of an object or the hearing of a sound, may cause rotation. A slight condition of weakness of the nerve-centres is expressed by flexion.

In the brain condition whose mental action is called "shame," and in the state of mental abstraction, the lessened kinetic function of the brain leads to head-drooping. Rotation, from the pathological conditions, hemiplegia and hemispasm, has been described in chap. vi. p. 105.

We all know that rotation of the head may result from the sight of an object, or the hearing of a sound. We will inquire first as to the effects of the sight of objects, or the direct rays of a luminous object, in causing rotation of the head. When the sight of an object is followed by the forehead being turned towards it, it is said to have attracted the attention of the man. The only connecting link between the man and the object that attracts his attention is the light reflected from its surface, or emitted from it, owing to its inherent luminosity; we may say, then, that it is the light from the object that causes the head to rotate the forehead towards the object or source of light. Thus, if a child is seated at table duly hungry, bringing a plate of food toward him will, as soon as the plate comes within his field of view, cause rotation towards the plate, and the plate of food will be said to have attracted his attention. If, however, the child is in a very irritable, cross, peevish state of mind (brain), rotation of the head may occur away from the plate of food, it may be repelled by it, not attracted. Here is a mode of expression; a similar stimulus at one time causes rotation towards the source of stimulation, in other conditions of the nerve-centres the head is rotated away from the stimulus.

The attraction and repulsion of the head as the effects of visual stimulus are very curious modes of expression. Probably the varying conditions of the nerve-system are often expressed by the effects of light and sound in causing rotation of the head. I think that this kind of attraction and repulsion can best be studied in the rotatory movements of the head and eyes.

Movements of the head frequently occur as the result of the spontaneous action of the nerve-centres. In infants there is frequent spontaneous movement of the head as of other parts. Head movements are often excessive in number in conditions of irritability.

Rotation of the head is often an expression of the mental state. This is specially seen when we contrast rotation with movements of flexion and extension.

Bulwer * speaks much of head movements as expressive of the states of the mind.

Rotation of the head is, according to the existing system of our English nerve-muscular organization, an expression of the mental state of "negation," or desire to deny. Flexion and extension express the mental state of acquiescence.

Now, sometimes we find a child so irritable that the predominant mental (brain) condition is such as to be all negation: the head rotates at all propositions—the proposal to go to bed, the proposal to eat, the proposal to walk, etc. The head rotation to all word sounds is a part of the expression of mental irritation or, rather, irritable condition of the mental centres.

Conversely, a child's state of frequent head flexion and extension to the stimulus of word sounds, indicates the mental state of ready acquiescence, or good humour.

It appears, then, that the condition of the brain in the mental state of "ready acquiescence" is one where symmetry of movement prevails, and the mental condition of "negation" is expressed by asymmetrical brain action.

Certain *postures of the head* are expressive of

* Bulwer, *op. cit.*

brain conditions, or emotions. In girls and in young people convalescent from chorea, etc., we often see inclination of the head, with slight rotation to the same side, and flexion. I have often observed, as a coincident hand posture, that we have in the free upper extremity the nervous hand on the side opposite to that of inclination and rotation. Very commonly in a weak child when told to hold out the hands to the front, we see a left nervous hand, with head flexion, and rotation and inclination of the right. In such a case the left nervous hand is the expression of the weakness and irritability of the right side of the brain, the head flexion shows weakness, and its rotation and inclination to the right express that the muscles on the left side of the neck are weaker than those on the right, giving another indication of the weak state of the right group of nerve-centres.

In studying postures of the head we have to consider some curious effects of gravity, explain them how we may. These principles affect, also, postures of the trunk, but I hardly think that they come into play in explanation of postures of other parts.

In a man with full health and strength the head is held erect. If the head falls, say, to the left shoulder, as the outcome of a temporary relaxation of the muscles of the right side, these muscles will be stretched, being on the convex side of the neck. This is the effect of gravity acting on the skull. This stretching of the right muscles transmits such an impulse to the nerve-centres as stimulates the right muscles, and they pull up the head erect again. If the man's head falls to his left shoulder while he is asleep, the nerve-centres for the right side of the neck are *not* stimulated by the pull upon the muscles sufficiently to make them contract. Here we appear to have a case of reflex action which is lessened by sleep, and is therefore an expression of the brain condition sleepiness. The same principles apply to falling of the head forward in flexion.

It is useless to our purposes to say that consciousness is a factor in the case, because we want to lead up to an enumeration of the signs or expression of consciousness. We want to show that the lessening of this reflex from the action of gravity is an expression of the brain condition which produces consciousness. I am not aware of any phenomena concerning the upper extremity that are exactly analogous to what has been just described concerning the head.

The head as a member of the body moved by the nerve-centres is usually free or disengaged, but it is not always so. It is not very free to afford expression to the emotions when a heavy load is being carried on the head. It may also be engaged, and not free, as the result of organic conditions, such as the fixed position of the head in an attack of spasmodic asthma.

Applying the principle of contrast of the action of small and large muscles, we should have to compare movements of the eyes or facial muscles with movements of the head. We may also compare movements of the hands and head. Extension of the head is the result of the action of a weaker piece of nerve-muscular mechanism than flexion; extension of the hands at wrists and knuckles often occurs with head extension, as in the expression of astonishment.

Movements of Lower Jaw.—The lower jaw is jointed on to the skull; in the natural upright position of the body gravity tends to make the jaw fall. Two pairs of muscles, the masseter, and the temporal, raise the jaw, and keep the teeth in contact; these muscles are both supplied by the motor division of the fifth pair of nerves of the brain. The masseter muscle arising from the cheek bone (the malar and zygoma) is inserted into the angle of the lower jaw; the temporal muscle arises from the fossa of that name, and is seen swelling out and may be felt as a hard mass in the temple when the jaws are tightly clenched. The jaw is depressed by a group of muscles supplied by the ninth pair of nerves of the brain. The mouth is often opened when the jaw is depressed, always when the jaws are opened widely, but the lips can be parted while the jaws are still closed, showing the teeth. The jaw is depressed and the mouth opened in the expression of astonishment.*

The jaw often drops somewhat as a passive movement, owing to relaxation of the muscles which support it. It is strongly depressed in yawning, occasionally so much so as to cause dislocation; this is owing to strong extensor depression of the bone.

* Darwin, "Expression of the Emotions," p. 280

Depression of the jaw elongates the face.

We will now look at some examples of expression in the permanent and unchanging conditions of the head—size, form, conditions of development. There are certain defective developmental conditions seen about the head which are not uncommon accompaniments of defect of brain, and which therefore express, though with a considerable degree of uncertainty, the probability that the brain may be defective in its powers.* The following head peculiarities are pointed out by Dr. Shuttleworth: "Congenital idiots have often peculiarities of conformation of head or feature which bear the impress of original defect. A shelving forehead, a flattened occiput, a notable proportion as to size between head and body as in microcephally (small-headed idiots), a nut-shaped forehead as in scaphocephally, marked obliquity of the supra-orbital ridges, absence of the nasal bones, and a lamboid (Λ-shapen) vaulting of the palate, are some of the cranial peculiarities associated with *congenital* forms of idiocy. We shall see also that the osseous irregularities in the form of the hands and feet, fissures and hypertrophy of the papillæ of the tongue, and a coarse, sometimes branny epidermis, are diagnostic marks of one congenital variety. The external ear is often unshapely, planted low down and comparatively far back; the angle of the jaw is more obtuse in congenital idiots than in normal children."

Summary.—In this chapter certain modes are

* See laws of proportional development, chap. xviii.

suggested for defining the postures and movements of the head, and the movements and combinations of movement possible to the head are explained. It is shown that flexion and extension are the only symmetrical movements of the head.

The action of light and the sight of objects in causing head rotation present interesting examples of expression, and some of the varying conditions of mentation are expressed by variations in such movements under the same stimulus on different occasions.

Head rotation is a sign of negation. Semiflexion of the head with rotation and inclination to the same side indicates weakness; this is often seen when the free left hand is in the nervous posture, the rotation and inclination then being to the right.

Gravity exerts a marked influence upon the postures of the head, especially in weakness and in sleep. The head is usually free, and therefore expressive of brain conditions.

The principle of the contrast of the action of small parts with larger parts, is illustrated by comparing the head movements with those of the face and eyes. Extension of the head and hands occurs coincidentally in astonishment.

Movements of the lower jaw depend upon two pairs of brain nerves. Depression of the jaw is spasmodic in yawning; it may be due to gravity.

Certain forms of the head are indicative of defects of brain, and are commonly seen in idiots.

CHAPTER XI.

EXPRESSION IN THE HUMAN FACE.

The face as an index of the mind—Definition of the face; its structure—Facial muscles and their nerve supply; the sympathetic nerve—Form, colour, and mobile conditions of the face—Direct expression of brain action in the face; expression by coincident development—Action of the facial muscles—Method of analyzing a face: the upper, middle, and lower zones; symmetry; analysis of the expression of anxiety—Expression by trophic signs, skin—The intellectual, and the vulgar face—The necessity of considering nerve-muscular signs as well as permanent conditions—Faces of idiots—Nutrition of the face; a dull and a bright face—What may be seen in a man's face—Impressions of previous movements of the face—Mental suffering compared with bodily suffering—Asymmetrical expressions; winking, snarling—The long face, due to paralysis of the nerve on one or both sides: due to mental states; facial palsy from brain disease—Intellectuality of facial movements—The face in fatigue; the expression of headache—The disengaged face free for mental expression—Cases of expression from brain disease—Conflict of muscles in the face.

THE human face has been often described as an index to the mind, and in most treatises on expression the passive and the changing conditions of the countenance are largely discussed. It is, then, unnecessary that I should say anything to prove that this subject is one of wide interest; but it must be treated of here according to the general purposes of the volume, and according to the principles and modes of expression already enunciated. We are here dealing only with material things and with the action of physical forces.

The human face may be spoken of as that part of the head which lies in front of a vertical plane passing just anterior to the ears. This includes the anterior portion of the skull with the soft parts attached thereto. These soft parts consist of the facial muscles which move the features of the face, and certain muscles of mastication; the interstices between the muscles are filled in with fat. Skin covers the whole face; it is in part adherent to the subjacent muscles, and is moved by them. The skin is continuous with the mucous membrane at the openings for the mouth, nose, and eyes.

The facial muscles proper, the muscles of expression, are supplied with motor stimulus by branches of the facial nerve, the muscles of mastication being supplied by the motor division of the fifth pair of brain nerves. Vessels supply blood to all these parts and to the skin. Branches of the sympathetic nerve supply the muscular walls of the small arteries, and by their action control the amount of blood supply; palsy of the sympathetic nerve on one side leads to flushing of that half of the face; thus the mobile colour of the face is largely controlled by the sympathetic nerve.

When we look at a human face we may observe its form, colour, and conditions of mobility. The general form and outline of a face is largely determined by the shape of the skull beneath. There is probably more direct expression seen in the face than in any other part of the body. The face is an index of the brain: the mobile conditions of the face are so many direct expressions of the brain condition; especially are those fine shades of variation expressive which accompany emotions and mentation.

We also see in the face many examples of expression by uniform coincident development;* this is an empirical form of expression, in which the condition of development of the parts we can see indicates the probable state of the brain which we cannot see. These different modes of expression must be considered separately.

In studying such a difficult and complex subject as the direct expression of the face, certain methods must be followed, and the different facial expressions that have been observed must be analyzed and described. The principal movements of the facial muscles are:—

1. Dilatation and contraction of the facial foramina—the openings of the eyes, nose, and mouth.

2. Elevation and depression of parts, as the eyebrows, the angles of the mouth, etc.

3. Retraction and drawing forward of parts, as in grinning and screwing up the mouth, corrugation of the forehead.

Either side of the face can move separately;

* See chap. xviii.

hence the necessity, in analyzing a facial expression, to observe whether it be symmetrical.

An expression may affect the face principally in the upper, middle, or lower portions, and it may appear more on one side than on the other; hence, in analyzing a face, each half and each region must be examined separately. I have found the following method convenient for making an analysis.

To examine a face, hold a sheet of paper in front of it, with one edge vertical, and opposite the middle of the face; either half of the face can then be covered in turn while the other half is examined. Again, the face may be divided into three zones, or horizontal areas,—the upper, middle, and lower. To observe each zone in turn, hold the sheet of paper with one margin horizontal, leaving the forehead above the eyebrows uncovered,—this shows the upper zone; then view only that part of the face which is below the lower margin of the orbits, showing the mouth, the greater part of the cheeks, and the openings of the nose,—this is the lower zone; lastly, the middle zone may be demonstrated alone by holding the horizontal margin of one sheet of paper so as to cover all above the eyebrows, and another sheet of paper so as to cover all below the orbits, thus leaving to view the eyebrows, the eyelids, and eyeballs, with the bridge of the nose.

By these methods we can easily examine for symmetry in a face, both as regards form and action; and we can, at the same time, also observe any special nerve-muscular conditions in any particular zone.

To illustrate by a supposed case, say that our common experience tells us that a certain man presents a facial expression of "mental anxiety," and on personal inquiry he acknowledges that he is suffering from causes producing "mental anxiety." On making a physical analysis of the expression by the methods suggested, we find that the expression is equal on either side—therefore the expression is symmetrical; we see the special signs of anxiety more when observing the upper zone than when looking at the middle and lower parts—hence the expression is symmetrical and principally located in the forehead, or upper zone. The nerve-muscular signs of the brain condition whose mental action is "anxiety" are localized in the frontal region, and these signs are direct expression. In thus studying a face we look to the nerve-muscular condition of the various regions, and observe the effects of the kinetic action of the brain.

We may next inquire what the trophic conditions of the face teach us. We look for those signs in the face which experience teaches us are commonly associated with certain coincident conditions of brain development—conditions of the skull or brain case, its form and size, the form of the forehead, etc. The structure of the skin of the face, whether it be fine and thin or coarse and thick; the features of the face—the lips, cheeks, nose, and the size and proportions of the mouth—demand attention.

In looking at different types of faces, we are at once struck with the fact that the passive appearance of some expresses intellectuality, while others are marked by inborn vulgarity, apart from any special expression by nerve-muscular action. Elements contributing to the low vulgar type are a large prominent under jaw, thick lips, a thick immobile make of skin, etc. Here the face is more fitted to bear exposure than to show fine nerve-muscular action, and the expression as to the mode in which the brain will probably act is due to the empirical fact that this type of face does coexist with that state of brain development which makes the individual tend to vulgarity in mentation and in action. This is only an empirical statement dependent upon the results of observation, and as such is likely not to be true in a particular case; indeed, it is often an untrustworthy sign. Such observations need to be corrected, by noting many nerve-muscular signs before we can determine the character of the individual. A somewhat plain or even vulgar-looking face may receive an intellectual expression when the brain is in action from mental work; other faces are most pleasing when passive, but when in action show a poverty in their nerve-muscular expression indicative of low organization of the brain-centres. Speaking of idiots, Dr. Langdon Down remarks (*Obstetrical Transactions*, vol. xxii.) that "the prognosis is, contrary to what is so often thought, inversely as the child is comely, fair to look upon, and winsome."

The condition of nutrition of the tissues of the face is an important index of the general nutrition of the body, and its different organs. A slight amount of malnutrition makes the face looks dull.

This, I believe, is due to slight absorption or shrinking of the fat of the face, leading to very fine wrinkling of the skin, which then reflects light in such a manner as to look dull. The dull skin looks bright, if it be stretched; a temporary afflux of blood accompanying a state of mental excitement often makes the face look bright, because it swells up the tissues and stretches the skin, thus removing the wrinkles and the cause of dulness of the countenance.

If we examine the face of a man, say, thirty years old, we may observe—

(1) The passive conditions resulting from heredity.

(2) The present trophic condition, the state of nutrition.

(3) The marks or permanent impressions made by the nerve-muscular actions during thirty years.

(4) The nerve-muscular condition at the time of observation.

This shows the necessity for balancing the different modes of expression, and, while observing the total expression, assigning the due value to each of its factors. We have spoken of the impressions made by the nerve-muscular actions of preceding years; if the muscular action in the face due to the condition of brain which accompanies mental anxiety has been often repeated during many years, it leaves permanent marks upon the tissues of the face.

The expression of mental anxiety may be contrasted with that of bodily suffering. Mental anxiety is expressed mainly in the upper zone of the face; contraction of the corrugators makes vertical furrows between the eyebrows. In the facial expression of pain originating in the body or limbs, we see the signs mainly in the lower zone; the angles of the mouth are drawn down. In the more animal-like causes of pain of mind—as the loss of a child, wounding the maternal instinct—it is the angles of the mouth that are depressed. Some years after the loss of the child a reference to it causes corrugation. The memory of the child has become idealized; the suffering is now more mental, less animal-like. A man of, say, forty years of age may present the permanent marks of anxiety impressed upon the skin of his forehead; this sign of anxiety observed may be due neither to an early original make of body, nor to the present action of nerve-centres, but what we observe is the permanent impression left by the frequent repetition of the same nerve-muscular signs.

If we see the expression of mental anxiety in the upper facial zone, and bodily suffering expressed in the lower zone, what do these facts teach us? These nerve-muscular signs are the direct outcome of brain action; so are the coincident mental anxiety and pain of bodily suffering. We conclude, then, that the brain centres which cause mental anxiety, and its expression in the upper facial zone, are different from the centres concerned in physical suffering and its expression in the-lower zone. Both these forms of expression are symmetrical, or alike on both sides.

I know of only two forms of facial expression that are not symmetrical—snarling and winking; and certainly these asymmetrical expressions are not the most intellectual. In snarling the canine tooth on one side is uncovered by the action of the levator of the upper lip on that side. This, of course, directly indicates the action of the corresponding nerve-centre on one side only. In one-eyed winking the orbicular muscle acts on one side only. This also shows asymmetry of nerve-action.

Suppose a man in "a quiet frame of mind;" we should see a general good balance of the muscles and nerve-centres, presenting symmetry of the facial muscles. The man then voluntarily winks with one eye or sneers unconsciously. Here an asymmetrical action replaces the quiet symmetry of the features. A part of the expression of passion may be asymmetry of action of the nerve-centres.

Writers of fiction, if also observers of nature, sometimes describe a man who has just received some depressing news—say, a heavy bill—as presenting a very long face. Can we, from actual observation and analysis, say that a long face is a mode of expression? The longest face I ever saw was in a patient with complete paralysis of both sides of the face of over twenty years' standing. The palsy was the result of disease of either ear, which commenced in childhood, and destroyed the facial nerve on each side. In this poor man, the face being constantly acted on by gravity, the tissues had fallen, the muscles had become stretched and permanently elongated, the skin of the chin had fallen far below the chin-bone, and literally the face was a very long one. In a case where one side of the face is paralyzed by destruction of its motor nerve, the paralyzed side after a time drops under the action of gravity, and it is easy to compare the two sides of the face. In such a case I have demonstrated by measurement that the paralyzed side may be three-quarters of an inch longer than the other side when in action. The actual length of the face can then be increased if the muscles are paralyzed; so also if they be relaxed from want of nerve-force coming

Fig. 27.—Complete Paralysis of the Right Side of the Face. Face drawn to the left side.

to the muscles. A face that is long, owing to nerve-muscular conditions, may be a direct expression of the brain condition; a relaxed condition of the facial muscles, allowing the face to fall and lengthen, is the outcome of feeble nerve-currents coming down from its nerve-centres. A long face may, then, indicate weakened brain force, and this often accompanies the mental condition following from a sudden disappointment. Another factor in producing a long face as a temporary condition, is the falling of the lower jaw.

Several modes of facial expression affect specially the lower zone. This is the region of the face that is the most weakened by brain disease.* A slight amount of brain damage in one hemisphere will usually produce, for a time at least, a certain amount of weakness on the opposite side of the face, in its inferior part. This facial weakness is easily demonstrated by making the patient show his teeth, or smile. The face when in action is asymmetrical, and the muscles in the lower zone on one side about the mouth act very indifferently; the grove running from the side of the nose to the mouth on that side is almost lost; that angle of the mouth falls lower than the other. No such asymmetry is seen in the upper and middle facial zones. Now, the muscles of this region are those most commonly seen in spontaneous action in imbeciles; it is these muscles that work so awkwardly in nervous one-sided grinning. Seeing, then, that the muscles in different facial regions are differently affected by brain conditions, the question presents itself, can we say that one region of the face represents intellectuality more than another?

Some years ago I attempted to determine this problem on the negative side by observing the conditions of the face in fifteen idiots. In conjunction with Dr. Fletcher Beach, of Darenth Asylum, I analyzed each face according to the following form:—

* See Fig. 11, p. 109.

General Muscular Condition.—The action, or relaxation, of the muscles of the limbs and body generally were noted.

Face.—Facial aspect, and muscles in action or relaxation.

Upper Zone.—Frontal region, occipito-frontalis and corrugator.

Middle Zone.—Eyelids and orbicularis oculi.

Lower Zone.—Mouth muscles; muscles of nose; cheeks.

Summarizing the results, we obtained figures showing the frequency with which these muscles respectively come into spontaneous action in a meaningless manner. This is perhaps some indication of the degree of their intellectual representation. Thus the corrugator and orbicularis oculi were much less frequently thrown into meaningless action than the occipito-frontalis and zygomatic, and probably the former are much more expressive of intellectuality than the latter. Again, applying direct observation to the other side of the question, and noting which muscles are most frequently put in action in the faces of intellectual people in the expression of their mental states, I think that we see intellectuality most commonly expressed in the frontal and middle zones, while grinning, yawning, and the meaningless smile are seen in the lower facial zone, that region which is the most readily affected by brain disease. It is not uncommon for nervous people frequently to produce a peculiar, awkward, grinning expression, owing to unequal action of the

THE FACE IN EXHAUSTION.

muscles on the two sides of the mouth. Such asymmetry produces a very unpleasing effect.

A general condition of fatigue is expressed in the face by a relaxed, toneless condition of the muscles; the face elongates or falls slightly, and the variation of the mobile expression, the play of the features, is lost.

A special sign of exhaustion is seen in those who

Fig. 28.—Imbecile; head well shapen and of fair size; he often smiled, thus moving parts around eyes and mouth.

Fig. 29.—Imbecile; same case quiescent.

have suffered habitually from recurrent headaches. When specially studying the faces of patients, the subjects of recurrent headaches, and analyzing them daily, my attention was particularly drawn to the middle zone. It is not uncommon to observe that an individual "looks as if he had a headache." Analyzing such faces, it soon became noticeable that there was a look of depression, heaviness, fullness about the eyes, especially about the under eyelid. It appeared that this expression must be due principally to the condition of the orbicularis palpebrarum muscle. There was no pitting on pressure, therefore obviously no dropsy, and when the face is dropsical this relaxed full appearance is not seen. Further, if the patient were made to laugh, the muscle became energized, and tucked the skin in well against the eyeball; thus the expression of headache was lost for a time. Specially observing the orbicular muscle and the parts adjacent, there seemed to be a loss of tone in it; there was an appearance of fullness and flabbiness; the skin hung too loose, with an increase of the number of folds; and, in place of falling against the lower eyelid neatly, as a convex surface, it fell more or less in a plane from the ciliary margin to the lower margin of the orbit. This condition is often best seen by looking at the face in profile. It was often seen when the skin was healthy and elastic; especially in children, and it could be completely removed by energizing the child. It passed away, in many cases, as health improved. It is not suggested that the nerve-muscular condition indicates only the states of brain producing headache; it may occur in other states of nerve-depression.

We have now considered several distinct modes of facial expression dependent upon nerve-muscular conditions—difference in the tone and conditions of contraction of the different facial muscles. These conditions are a direct expression of the brain action, because they result solely from brain action.

The facial muscles are indices of the brain action for the same reason that postures or movements of the hands are expressive. It is only when the facial muscles are free and disengaged that they can receive, and be acted upon by, the finer nerve-currents coming to them from the brain. It is, then, the disengaged face in which we see the most expression of that action of the brain which produces the emotions and mentation.

Observing a man while his facial muscles are performing the movements necessary to eating, or while a strong light is shining in his eyes, does not afford a good opportunity for observing the effect of mental action in producing mobile conditions of expression. While the facial muscles are passive, and not under the influence of any strong stimulus, they are much more impressionable to nerve-currents coming down from the centres of mentation. When strongly acted on by some reflex stimulus, they are less impressionable to the slight ever-changing currents coming from the centres of mentation.

When the orbicularis muscles are contracted from a light shining in the eyes, when the semi-reflex actions of eating are going on, or while whistling, the facial muscles are less expressive of the condition of the mind. Conditions of disease of the brain may render the movements of the face inexpressive of the state of the brain as regards its function of mentation.

The following case should probably be classed as one of athetosis of the face, associated with *petit mal*.

John Clark came under my observation May, 1878; he was then twelve years of age. The complaint made was that his hands twitched, his face worked much, and that at times he was quite silly.

He was a well-grown lad, of good complexion; his features presented a constrained appearance, and there was much movement of the face. The right hand twitched, but not violently. As he stood up there was some irregular muscular twitching all over the body. He spoke well, and was not himself conscious of the twitching of the face.

The movements of the face were very striking, and were carefully observed. They were principally confined to the parts about the eyes, nostrils, and mouth.

In July, when he was an in-patient at the East London Children's Hospital, I took the following description :—

Face: complexion somewhat anæmic, mucous membranes a little pale, not markedly so. There is considerable movement of the muscles of the face; he frequently closes his eyes, next draws up the angle of the mouth on the same side, then contracts his eyebrows (corrugators), next he elevates them. He appears quite unaware that he is making grimaces. The inner portions of the occipito-frontales (grief muscles) frequently contract in excess of other portions of these muscles; no movements of the hair or scalp seen.

The orbicularis oculi is frequently contracted, as indicated by the furrows around the eye, and the movements of the lower lid, while at the same time the eye is sufficiently open to allow of the pupil being distinctly seen (coincident action of the orbicularis and levator palpebræ ?).

The eyes are much moved in a restless manner, and are frequently turned upwards when the eyelids are being closed. There is never any strabismus. Fundus oculi normal. The tongue is protruded at request, and kept out, and there is much irregular movement of its muscular structure.

As the patient stands upright with his heels together, there is scarcely any movement of the body. When he holds out his hands, and spreads his fingers, they are kept steady, but very slight abductor movements and almost inappreciable flexor-extensor movements of the fingers are seen; there are also slight movements of the toes. He walks well, with a steady gait, but slight irregular swaying movement of the head.

His heart and lungs appeared normal.

The following was the history of the case: He had always been strong and well till a year and a half or two years previous to my seeing him. He never had convulsions; he went to school when five years old, was bright and intelligent, and did as well as other boys till the autumn of 1876; never any complaint of pain till this time. About that date it was noticed that he made contortions of his face, and frequently nodded his head; then, later on, the fingers of the right hand began to work. After some months it was noticed that his manner was changed; he became forgetful and vacant—at times as if silly.

When sent on errands he went to wrong shops. He appeared at times to lose himself. One day he was found in the middle of the road at a distance from home, and could not say how he came there or where he lived. He became a tiresome boy, and at times passionate. If beaten he became worse, as if he had lost his reason entirely. In hospital he conversed well, and no mental failure was noticed.

Sleep was sound and refreshing.

As to the inheritance, the father and mother, and the families of which they were members, appeared healthy. The following is the account of the collateral members of the family:—

(1) Male 25 years, healthy.
(2) Male 22   ,,   ,,
(3) 5 children died in infancy.
(4) Patient, aged 12 years.
(5) Male 9, very healthy.
(6) Female 7, not strong.

The possibility of some defective inheritance is the only probable cause of the boy's illness.

When seen, January, 1879, he went to school, was much less absent and forgetful; movements of face were less, principally consisting in contractions of both occipito-frontales muscles. No fits or attacks.

Now (August, 1880) he goes to school; seldom or never falls into "the lost condition." His face is peculiar, and somewhat expressionless, but without pathological movement except in the forehead. As he talks the skin of the forehead is frequently thrown into fine long transverse wrinkles by a slow movement.

Hemiplegia: involuntary movements of arm and face.

Arthur T——, aged ten years, came under my observation (May, 1880) as the subject of a chronic left hemiplegic affection. His general health had been good. His left upper extremity was wanting in muscular voluntary power and in muscular sense; certain involuntary movements of the hands and of the fingers were also observed. I now collect the notes describing the nerve-muscular condition of the hand. The left hand, when held out, assumed on the average the "nervous position;" not so the right, which was natural. A certain amount of involuntary movement of the fingers was seen; the middle finger moved the most; during the examination, and at other times when his fingers were curling up against his will, he would frequently use the right hand to strengthen the fingers of the left. As to the kind of movement, it certainly was not voluntary, and on several occasions a similar movement was seen repeated. The dynamometer showed the power of the right hand as 47, against 36·5 on the left side.

I had often noticed that the boy looked much distressed when I examined the left arm, and this appeared not surprising, considering that, with good bodily health, he was losing the use of the left arm, and at the same time there was family trouble from the father being out of work. Looking at the lad's face, one observed distinct over-muscular action in the upper zone, causing in the middle of the forehead transverse and vertical furrows, an appearance commonly resulting from the condition of brain associated with grief. For a long time I was deceived by this face, and thought the boy depressed, but later, on cross-questioning him, and inquiring of his mother, he appeared not only unconscious of these movements but not to feel any mental distress. I was therefore compelled to regard the facial movements, which were suggestive of the expression of distress, as analogous to those of the fingers, which more resembled athetosis than any other phase of involuntary movement. The movements of the fingers were slow, involuntary, and unconscious; no child in chorea puts up his hand to straighten out the fingers which have curled up on the other hand. There were no general twitching movements, and those that did exist were not twitching in character. These points appear to indicate an analogy to athetosis, rather than to chorea, both in the hand and in the face.

Throughout the mechanism of the human body the muscles are arranged in groups, acting antagonistically to one another. In the limbs flexors oppose the extensors, pronators oppose the supinators, etc.; likewise in the face the muscles, which contract the openings oppose the muscles which dilate them. These opposing groups of muscles may be energized by nerve-currents, either separately or together, or more commonly they are stimulated to contract in unequal degree, and so the muscles which are stronger, or the most stimulated, produce the actual movement. If the flexors are the most stimulated,

CONFLICT OF EXPRESSIONS IN THE FACE. 213

flexion results. If strong and unequal nerve-currents are sent to the opposing muscles, a quivering or tremor of the part moved by the muscles may result. Such action is often seen about the muscles of the mouth under the influence of "conflicting emotions." Suppose a child has hurt his finger but is trying hard not to cry, we shall see the

Fig. 30.—Paralysis Agitans, in advanced stage. Face almost expressionless, with loss of all the fine adjustments of expression, it presented one dull monotony of appearance.

muscles of the mouth quiver, until finally the effect of the injury to the finger acting upon the nerve-centres is too strong for the action of his will, and the angles of the mouth are depressed and the outbreak of sobbing follows. The conflicting emotions, or the conflicting nerve-currents acting upon the muscles cause the expression about the mouth.

CHAPTER XII.

EXPRESSION IN THE EYES.

The eyeballs: their position and the mechanism for their movement—Iris, a muscular apparatus; its nerve supply—The pupil contracted by light, and accommodation for near vision; its reflex dilatation, and its variation in conditions of emotion and on brain stimulation—Mechanism of the eyelids—Importance of distinguishing expression by the eye and the parts around—Movements of the eyes—Loss of associated movements of eyes under chloroform, and in deep sleep—Movements of eyes from brain stimulation—Horizontal movements most common—Attraction and repulsion of the eyes by sight of an object—Spontaneous movements—Eyes free or disengaged—Mental states expressed by attraction or repulsion of the eyes—Horizontal and vertical movements contrasted—Intellectuality of upward movements.

MOST people, if questioned, would say that in the human face there is great expression in the eyes; that the eyes are very expressive features, and that the soul and mind shine out in the eyes, etc. In conversational language, which is not always quite precise, the term "eye," as a feature of the face, is used somewhat loosely; but it is necessary in this method of study to differentiate carefully between expression seen in the eyeball, and expression in the parts that surround the eyeball—the eyelids, the eyebrow, etc.

The eyeballs lie in their sockets, the orbits of the skull, resting among the fatty tissue which supports them. If that fat be diminished in quantity, the eyeball sinks further into the orbit; if the fat becomes congested and swollen up, it protrudes the eyeball somewhat. The movements of the eyeball are effected by small muscles attached to the eye and arising from the wall of the orbit; these small muscles external to the eyeball are supplied with motor force by the third, fourth, and sixth pairs of nerves of the brain. The iris, or coloured portion of the eye, is a muscular curtain, with an aperture in its centre called the pupil. The movements of this muscular apparatus, the iris, are under the control of the sympathetic and third cranial nerves. Stimulation of the sympathetic is followed by dilatation of the pupil; stimulation of the third nerve, or motor oculi, is followed by contraction; conversely, palsy of the sympathetic leads to a small pupil, and palsy of the third nerve to a large pupil.

We have seen, then, that a very brief description shows how the size of the pupil may be increased or diminished by nerve-currents acting upon the muscular apparatus of the iris. Such changes are among the principal means of expression in the eyeball. The varying size of the pupil is a mode of expression.

(1) Light falling upon the eye enters the pupil and falls upon the retina, or nervous impressionable layer of the back of the eye; this sends on a current to the corpora quadrigemina of the brain, and is

reflected by the third nerve, causing contraction of the pupil. Light falling upon the eye contracts the pupil.

(2) If an observer stand in front of a man who is looking out of window at a distant object, he sees his pupils of a certain size. If he now requests the subject to look at a finger held three inches from his eyes, a movement of the iris is seen, the pupil contracts, at the same time the eyes converge so as to direct their axes towards the finger looked at. This contraction of the pupil is due to accommodation of the eye for near vision.

(3) The size of the pupil may vary as the result of a mechanical stimulus applied to some distant part of the body. Thus, gently stroking the palm of the hand may cause slight dilatation of the pupils; they may also be observed to react to stimuli applied to the face and neck.*

(4) Certain so-called "emotional states" cause variations in the size of the pupil, that is, the changes in the size of the pupil are expressions of the emotions; the material change which produces the emotion produces at the same time a change in the iris. Ferrier † showed by direct experiment, that, in pigeons, irritation of the optic lobe on one side causes the opposite pupil to become intensely dilated.

A short account must now be given of the mechanism for the movements of the *eyelids*. The

---

* See Dr. Wilks "On the Pupil of Emotional States." "Brain," part xxi., 1883.

† Ferrier, *op. cit.*, p. 78.

lids are closed or drawn together by the action of the orbicular muscle of the eye. The upper lid is raised by the action of a special muscle, termed elevator of the upper eyelid; this acts in opposition to the orbicular muscle. The orbicular muscle, which closes the lids, is mainly supplied by the facial nerve; the elevator muscle of the upper lid is supplied by a branch of the third nerve, the motor oculi; so that different nerves are concerned in opening and in closing the eyelids. We shall refer to this again, in considering some points concerning the movements of the eyes and eyelids.

Much has been said by different authors about expression in the eyes, but many descriptions do not particularize as to whether the expression is seen in the eyeballs, or in the features of the face, the eyelids, and the parts around them.

I think there is more expression in the action of the muscles of the eyelids than in the changing conditions of the eyeball itself. If a man wear a mask, showing the eyes only and hiding the other features of the face, there is so little expression seen that it is impossible to recognize the individual thereby, as may be seen at a masked ball. It is the custom in some parts of Italy for men to beg in silence, wearing a loose garb, and a hood covering the face, with holes showing the eyeballs only, and the absence of expression is marked.

In looking to the modes of expression in the eyes, as in other parts of the body, the greatest number of signs are seen in the nerve-muscular conditions and in the movements. When the eyes are directed to the right hand or to the left, the axes of the eyeballs move horizontally, and remain parallel to one another. Both eyes may have their axes directed to the right hand or to the left, but the parallelism of the axes is maintained. This co-ordination of the centres of movements is constant during health while awake. If, however, the patient be placed under the influence of chloroform, so as to produce perfect unconsciousness, the loss of associated movements of the eyes may be complete. If, in an adult completely anæsthetized with chloroform, the upper eyelids be gently raised, the pupils will be seen minutely contracted, often to a pin-point, the eyes at the same time having lost the parallelism of their axes. One eye may move inwards or outwards, while the other remains quiet, or moves in a different direction or at a different pace, thus causing a temporary and varying squint. Usually these movements are mostly in the horizontal plane; less commonly the eyes assume a different level, one being in the horizontal plane, while the other is turned downwards. These movements I have frequently demonstrated in the healthy subject, and have seen that though they occurred in this unconnected way during the coma, they regained their parallelism when the effect of the chloroform passed off, the pupil expanding at the same moment. In an infant in profound sleep the same loss of association of the movements of the eyes occurs; it may also be seen in young and feeble infants when sucking at the bottle. These points, then, will serve us as indications of sleep, coma, health, etc.

Ferrier * has shown that there exists in the convolutions of the frontal region of the brain a centre, which, when excited, causes both eyes to turn away from the side excited; if, on the contrary, this centre be destroyed instead of excited, the corresponding centre in the other half of the brain, acting unopposed, turns both eyes towards the side of lesion. This physiological fact suggests the importance of studying lateral movements of the eyes as a sign or expression of brain action.

In health the greater number of ordinary movements of the eyes are probably in the horizontal plane of the axes of the orbits; the same is the case with the spontaneous movements of the eyes in a patient under the influence of chloroform. The horizontal movements of the eyes do not involve as many separate muscles as the vertical movements, inasmuch as they do not involve movements of the eyelids.

Many of the lateral eye movements are due to the sight of objects. "The sight of a flower" is said to attract a child's attention if he turn his eyes towards the flower presented. Here is an interesting case of eye movements expressing that the "attention" is "attracted." If movement results from the sight of the flower, the flower has something to do with the movement. There is no material, structural connection between the flower and the child's eyes; it is light reflected by the flower which passes to the child's eye and stimulates its movements. It is not every visual stimulus

* Ferrier, *op. cit.*, p. 229.

that causes the eyes to be directed towards the object; the sight of some objects repels the eyes. The sight of an object may at one time attract, at another repel, the eyes. We see, then, that the terms "attract," "repel," as here used, are purely conventional; there is no such thing as attraction in the sense in which the word is applied in speaking of gravity, or electricity.

As an illustration of the movements of eyes under the effect of the sight of objects, I may refer to observations I have often made upon people during railway travelling. It is easy to make observations upon the movements of the eyes of a fellow-passenger sitting opposite and looking out of window. When the line of rail is running on an embankment, there being no tall objects near the train, the distant objects remain long in view, and as the eyes are attracted to one object in the distance after another, the eyes move horizontally, but slowly, their axes being directed to one object till it passes beyond the field of vision, then moving towards another object. When the train passes through a cutting all objects within view are near, and remain within sight but a fraction of a second, so that as one object after another attracts the attention the eyes move rapidly—I think often over a hundred times a minute.

Having considered these preliminary matters with regard to eye movements, our attention must now be directed to the principles and modes of expression as seen in the eyes. In a young infant the eyes are constantly moved while it is awake;

we may conclude, then, that movements of the eyes, as of other parts, may be spontaneous, due to spontaneous action of the nerve-centres. Movements of the eyes are very expressive. Movements of the eyes, not coinciding with the special presentation of objects within the field of vision, are probably spontaneous; when the "sight of an object" is followed by the optic axis being directed towards the object and kept there, this shows the subject impressionable to the effects of light, and shows that the spontaneous movements can be inhibited by light.

Again, the iris is a muscular apparatus stimulated by certain nerves; its movements determine the size of the pupil. The size of the pupil often varies independently of any variation in the light falling upon it; it varies owing to changes occurring in the brain, particularly the changes which occur along with or produce the emotions.

Now, let us inquire whether there be evidence that the condition of the mind is expressed by movements of the eyes; in other words, whether those brain changes which produce mentation often cause synchronous movements of the eyes. Can anything be shown with regard to the intellectuality of eye movements? Eye movements are most expressive of mental conditions when the eyes are free or disengaged. If the optic axis has been directed towards an object by the stimulus of light reflected from that object (*i.e.* a visual stimulus), the centres which move the optic axis to its position remain under that stimulus, and this position of the eyes is maintained till that stimulus is removed, or some stronger stimulus excites the nerve-centres. While the eyes are thus fixed by the sight of an object, or other visual stimulus, they are not altogether free or disengaged to express emotions; their nerve-centres are engaged as much as those of a hand while digging.

A sufficiently strong visual stimulus may direct the eyes towards the object, or repel them; that is, direct them away from the object. In these varying effects of the sight of an object we have one of the most interesting modes of expression. A glass of water placed within the field of vision of a thirsty man causes his eyes to turn towards it; the glass of water, if placed within the field of vision of a man suffering from hydrophobia, causes repulsion of the eyes. Some explanation of these phenomena may be seen on referring to Ferrier's experiments.

When a patient is under the influence of chloroform, a number of facts show that he is less impressionable to outward stimuli. Touching the cornea does not excite reflex movement; speaking to the patient does not excite movement. The movements that do occur in this condition are then probably spontaneous (automatic). It has been already stated that in this condition spontaneous movements of the eyes are seen, and they are mostly in the horizontal plane. We have in these facts evidence that horizontal movements are more likely to be spontaneous than those in the vertical direction; this also makes it probable that as, when mentation is paralyzed, horizontal movements are most common, the vertical ones are more expressive of mental states. The horizontal movements of the eyes do not involve movements of the eyelids; these movements are less complicated than those in the vertical plane, and involve the action of fewer nerve-centres.

As to the mechanism of vertical movements, Dr. Gowers * has shown that the movements of the lower eyelid constitute a simpler problem than those of the upper lid. The lower lid follows the movements of the eyeball upwards and downwards, but not very closely. No muscular mechanism exists which can cause the downward movements of the lower lid; such movement is produced by the movement of the eyeball acting mechanically upon the lower lid. The upper eyelid possesses a more complex mechanism. The descent of the lower lid, or downward rotation of the globe, is not due to the contraction of a muscle, but simultaneously with the descent of the upper lid in the downward movement of the eyeball, there must be a relaxation of the elevator. In upward rotation of the eyeball, contraction of the elevator is associated with that of the superior rectus. The association of the elevator and superior rectus suggests that both are relaxed, or energized, in similar degree when the eyeball is moved upwards or downwards. Thus, in the upward vertical movements of the eyeballs, a more complex mechanism comes into play than with horizontal movements.

* *Med.-Chir. Trans.*, 1879.

In applying the principle of contrasting the movements of small parts with those of larger parts, it seems probable that we should compare movements of the eyes with those of the head. Probably there is more intellectuality expressed by movements of the eyes towards an object than when the head is turned instead of the eyes.

CHAPTER XIII.

EXPRESSION OF GENERAL CONDITIONS OF THE BRAIN AND OF THE EMOTIONS.

Expression of consciousness—Sleep—Fatigue—Exhaustion—Irritability—Nutrition—Rest—Activity—Impressionability—Expression of instinct and mentation—Expression of pain—The emotion of the beautiful.

VARIOUS modes of expression, or physical signs, the outcome of brain action, have been considered; we shall now discuss some of the brain conditions causing or accompanying the emotions, and certain general conditions of the man, noting how they are directly expressed.

It is proposed to speak of the following general conditions:—Consciousness; sleep; fatigue; exhaustion; irritability; nutrition; rest; activity; impressionability; expression of instinct and mentation; expression of pain, and of the emotion of the beautiful.

As to the direct expression of consciousness: *consciousness* itself in the abstract is beyond our sphere of observation, but the expression of consciousness is most important to our subject, and interesting to study. We are getting near a metaphysical question; to clear the ground, let me speak plainly. I have no idea what consciousness is, just as I have no knowledge what life is, or what mind is in the abstract; but we can study the expression of these unknowable things, consciousness, life, mind; we can study their physical manifestation. In studying an emotion or general condition, such as consciousness, by these methods, we must say that such and such are the physical signs produced by that condition of the brain which usually produces consciousness. The physical signs which we commonly say express consciousness are not actually the result of consciousness, but they are the outcome of the action of that nerve-mechanism which also affords consciousness; it is not the abstract "consciousness" that is expressed, it is the state of the nerve-mechanism which produces the expression.

I would not endeavour to adhere thus pertinaciously to our methods of procedure, were it not that it appears to me that many authors, in treating of problems in which "consciousness" becomes a factor, become lost in a maze of metaphysical inquiry. Now, pursuing our subject, the expression of consciousness, let us see what can be learnt by the direct observation of individuals acknowledged to be in the state of consciousness or unconsciousness, and contrast the signs observed. In conditions of consciousness there is great impressionability, susceptibility to impressions upon the organs of special sense, producing reflex movements; in states of profound unconsciousness, as a rule, no reflex movements follow impressions on the organs of sense.

We are considering, at present, expression in healthy subjects only, and the only normal condition of unconsciousness is sleep. We pass on by a natural step to study the objective signs of sleep. *Sleep* is a condition apt to follow from fatigue. The principal signs of sleep are, diminished impressionability, diminished power to resist the effects of gravity upon the postures of the body, absence of movements in the so-called voluntary muscles, alteration of the organic movements. Many details might be given, a few must suffice for want of space. The tone of the orbicular muscle of the eye is, in healthful sleep, sufficient to keep the eyelids closed; on awaking, the levator muscle of the upper lid preponderates, and the eyelid is raised. If during deep sleep the upper eyelids be raised, the pupils are seen minutely contracted, and in infants we usually see a complete loss of the associated movements of the eyes.[*] One eye may move upwards or outwards while the other remains quiet, or moves in a different direction, or at a different pace, thus causing a temporary squint. Usually these movements are confined to the horizontal plane of the axes of the orbits; at the moment of waking the pupils dilate, and co-ordination of the movements of the eyes is restored. I have frequently observed these facts in infants asleep in their mothers' arms. It is also

[*] See paper, *British Medical Journal*, March 10, 1877.

not uncommon during sleep to find the condition of muscular tone different on the two sides of the body. We commonly say there may be different depths of sleep; sleep may be full and complete, with loss of most forms of impressionability. Objective observation of a subject during apparent sleep does not necessarily give evidence as to whether there be any impressionability or not; it may be found that the outcome of impressions received is long delayed. Things said before a child apparently asleep may not produce any visible result at the time, but we may subsequently learn that it was impressionable to sound at the time of observation; the child may tell us what was said before it. Here we know the impressionability by its effects in subsequent speech (movements).

Fatigue is indicated by the slight amount of force expended in movement, and by the small number of movements. In the latter character we see some distinction between fatigue and irritability, in which condition there is often an excess of movement, and in particular an excess of speech. Fatigue and irritability often coexist. The voluntary movements are apt to become uncertain, as is illustrated by the change of handwriting when a man is tired. The free hand assumes the "straight extended posture with the thumb drooped," or the posture of the feeble hand. The head is often in an asymmetrical posture, and flexed. The direct effects of gravity determine the position of the body to a greater extent than in the condition of strength, hence the spine is bent. If this condition tends to pass on into sleep, this is expressed by the preponderance of the orbicular muscle over the levator of the upper lid, and the other signs of sleep supervene.

Exhaustion is an extreme condition of fatigue, in which movement (the kinetic function) is altogether lowered. The face becomes toneless, and devoid of fine mobile expression; the orbicularis palpebrarum is relaxed; the face may be lengthened from relaxation of its muscles; the ordinary movements of expression are not excited by the ordinary stimuli, and such movements as do occur are slow and laboured. A strong stimulus is required to induce the man to hold out his hands, and then the posture is " the feeble hand." Sighing and yawning are common; speech is slow, and the tone of the voice is altered; in some cases finger-twitching, especially of separate fingers, indicates extreme exhaustion and irritability.

Irritability is expressed in a man when a slight noise makes him start. This is a reflex movement in excess, a reflex action that does not occur in perfect health on so slight a stimulus. In irritability other stimuli besides sound may produce excessive reflex action. A touch upon the shoulder causes a sudden movement. Not only is the amount of reflex movement excessive, and out of proportion to the stimulus, but the kind of movement, or special series of movements, may differ from that usually following from the stimulus in health. A child three years of age, when irritable, may turn away his head from a familiar object, or from the sight of his food, and say, "No, no;" here the sight of the object, instead of causing a reflex movement of head, eyes, and hands towards the object, moves all from it. The irritability of the nerve-centres is indicated by movements in the opposite direction from that which the same stimulus would produce in health. Besides these reflex signs we find the voice altered—when spoken to he may answer sharply; the kinetic force generally is lessened and irregular in kind; twitching irregular movements are not uncommon. The nervous children described in chap. vii. often show marked signs of irritability. The spontaneous postures assumed are those of fatigue, with the addition of slight irregular twitching movements. If this condition lasts long trophic signs are usually seen, nutrition is lowered, and wasting occurs. It is not convenient here to discuss the relation of such trophic changes to the motor phenomena of irritability, but let me refer the reader again to chap. vii., p. 113, where some experiments upon the sensitive plant are described. Abnormal conditions in the body, particularly in the abdomen, may render the subject irritable; so may fever, gout, etc.

Nutrition is a matter of the highest importance. We must consider its expression. I am, however, unwilling to enter at any length upon the expression of nutrition in this volume, because the subject is so extensive and so important; in fact, the whole of this volume is preparatory to an attempt to gain knowledge upon the subjects nutrition, and development, as affected by external forces. The first point I wish to insist on is that nutrition may be expressed by (A) trophic, or material visible signs; (B) kinetic, or motor signs, the direct outcome of nutrition. The trophic signs are commonly known; the kinetic signs are perhaps too often passed over with scant notice. As evidence that kinetic signs or movements, and the results of movements, may express nutrition, let us examine a few examples.

(1) In an ill-nourished infant spontaneous movement is much lessened, or the child may lie almost motionless, instead of being constantly full of movement while awake. The return of spontaneous movement is a sign of the improved nutrition.

(2) In a man after a severe illness, such as a fever, the tone of the voice is usually altered so that we can no longer recognize the individual by his voice. This motor sign indicates, as well as the worn countenance, the man's lowered nutrition. Returning health is indicated by the patient "looking like himself" and "recovering his old voice."

(3) In a child seven years old emaciation and ill nutrition, indicated by loss of weight, may be accompanied by chorea or finger-twitching, which disappears when weight increases and nutrition is improved.

(4) A strong well-nourished man is less fidgety than a weak one.

Now as to the expression of nutrition by trophic signs. Proportional development is often an indication of conditions of nutrition. A seedling pea-plant, if kept in a room with deficient light, is not well nourished, and the ill nutrition is indicated by the small yellow leaves and the long white stem. That assimilation has not occurred during the life of the plant is demonstrated by the fact that the plant when dry weighs less than the seed from which it grew. Here ill nutrition is expressed by the relative growth of leaves and stem, the leaves being very small, the internodes very long. In children we often see growth for a time occur in height without lateral development; then the proportions of growth change, and the child fattens.

Rest is probably a condition of nutrition, leading to the signs of recreation indicated by subsequent activity. The most essential element in the expression of the condition rest is the subsequent activity, *i.e.* the sequential series of movements and reflexes. Take a case uncomplicated by sleep. During rest there is impressionability, which affords a distinguishing character between simple rest and sleep. Arising out of this we have the fact that in rest, uncomplicated by sleep, the eyelids usually remain open; that is, the levator palpebræ still receives a stronger impulse than the orbicular muscle.

One of the special characters of rest is the absence of movement, although impressionability is retained. Rest is usually preceded by fatigue, and it is followed by activity; the sequential signs of recreation and activity indicate that during the period in which movement was absent there was rest. Rest is expressed by the present signs of rest, followed by the signs of recreation and activity.

As a matter of interest it may be noted that forces, such as the sound of soothing music, may produce an inhibition of movements. Music may cause a man to keep quiet and rest. As a matter of speculation it seems to me probable that when the kinetic function of the nerve-mechanism is in abeyance the trophic action prevails, and more nutrition takes place. One reason for this view, among many others, is that a general good state of nutrition of the body promotes rest. Conditions of ill nutrition are often accompanied by signs of irritability.

In contradistinction to the state of rest we have activity. The condition of activity is indicated by actions, *i.e.* movements. In activity with strength, the movements are probably fewer in number than in the state of irritability, and the combinations and sequences of movement differ in the two conditions.

Healthy activity is indicated by a quick response of movement upon stimulation, that is to say, for example, that the movement follows quickly upon the sight of an object or on hearing a sound. If such movements are looked upon as reflex actions, the quick and ready answer is a reflex series of movements where the period of latency is short. This, of course, implies also that impressionability is good.

Impressionability, as a property of the nerve-mechanism, is especially interesting as being an essential factor among the properties of the brain necessary to mentation. In the first two chapters

I explained what is here meant by the term impressionability, and it was shown to be a property possessed by inorganic bodies as well as living beings; and impressionability of the brain is dwelt upon in the chapter on "Expression of the Mind." This is a very large subject, and there are many kinds of impressionability, such as to the effects of light, sound, etc.

A curious form of impressionability is the ready susceptibility to the formation or building up (trophic action) of new reflexes. A picture when first seen attracts the attention but little; the head and eyes do not with certainty and quickness turn towards the picture when it comes within the field of vision, but after a time the picture makes an impression upon the nerve-centres by the light it reflects, and the head and eyes then always turn towards it when it is in view. It is more difficult to understand how a series of movements spreading over some considerable period of time can follow from a single short impression. When the brain readily retains impressions it is said to be retentive, and *retentiveness* in the brain is very important to intellectuality.

The emotions are usually considered as the outcome of the mind, and before considering any of the expressions of mind it is convenient to consider what can be said about the expression of instinct.

Instinct is defined by Professor Bain* as untaught ability. "It is the name given to what can be done prior to experience or education; as suck-

* "Mental and Moral Science," 1872, p. 68.

ing in the child, walking on all fours in the newly dropped calf, pecking by the bird just emerged from its shell, the maternal attentions of animals generally."

In the larger edition of his works, p. 256, Professor Bain says, "Instinct is defined by being opposed to acquisition, education, or experience; we might express it as untaught ability to perform actions of all kinds, and more especially such as are necessary or useful to the animal. In it a living being possesses, at the moment of birth, powers of acting of the same nature as those subsequently conferred by experience and education." Again, later he lays some stress on "the primitive combined movements;" "the primitive arrangements for combining movements in aggregation or succession;" as examples, locomotion, and the aggregation of movements dependent upon the construction of the nerve-mechanism at birth seen in the movements of the eyes.

Walking is said to be due to instinct in the lower animals. Marey describes walking as a series of movements.

Sir Charles Bell,* speaking of alteration of the respiratory movements under the influence of passion, cites as an example the movements of alarm, the sudden and startled exertion of the hands and arms, etc., and he then adds, "Such combinations of the muscular actions are not left to the direction of the will, but are provided for in the original construction of the animal body."

* *Op. cit.*, p. 191.

It seems, then, to be acknowledged that what is called instinct is expressed by "movements in aggregation or in succession," due to the congenital or born condition of the nerve-system. If this be the meaning of what is understood by the term "instinct," we see that instinct may be expressed by combinations and sequences of movements, the direct outcome of the spontaneous action of the central nerve-mechanism as it exists at birth, or in addition by combinations and sequences of reflex actions, and these are in part dependent upon external stimulus. I think the chief notion implied by the term "instinct" is that the movements which express it, whether spontaneous or reflex, are the outcome of the inborn condition of the nerve-centres, and are not acquired; if they are shown to be acquired I suppose they would be called signs of "mind." We have, then, a more or less arbitrary definition, and a definition proves nothing; what needs demonstration is that instinct as thus defined does exist as a property or condition of things.

In our chapter on "Expression of Mind in the Infant" it is assumed that mentation is a function of the brain; we are now concerned with a certain class of mental phenomena called the emotions. The expressions of the emotions have been well described by many observers and authors, and I propose to analyze their descriptions instead of reattempting their task.

Pain is an emotion, and can be expressed in the body by certain physical signs. "Pain" in the abstract is outside the sphere of objective knowledge, and is incapable of investigation by physical means. In talking of pain in this work, I take some description of pain generally understood and adopted by us all, then observe cases where pain is present, and note and analyze the coincident physical signs. That is to say, we take the general, social description of the abstract emotion, and deal by our principles with the physical signs accompanying it. I do not wish here to say anything about the emotion itself, but only to deal with its expression in physical signs. To us the signs of emotion are physical signs, or indications of brain changes, and we grant that a physical sign can only be produced by a previously existing physical force. We do not say that an emotion is a physical force, but we do say that the physical sign which indicates the emotion is the direct outcome of a physical change, or force in the nerve-centres. We will not now discuss what may be the mutual relation of (*a*) the emotion ; (*b*) the physical sign which expresses it ; (*c*) the force or change in the nerve-centre which produces the nerve-muscular expression.

The emotion of beauty may be taken as another example. In this work we put aside the consideration of all subjective states, and abstract ideas. It is the expression of the sense of beauty that will engage our attention. I know not what "beauty" in the abstract is, but we can observe the physical signs which express that a man is affected by what we know is beautiful, or what we have reason to believe is beautiful to him. Here we touch upon a matter of interest and of importance to our subject.

Objects may be beautiful to one man that are not beautiful to another. The same object—say, a scientific instrument—is beautiful to one man, not to another; it affects them differently. Probably there is no property "beauty" resident, as a property, in the instrument; the sense of beauty, and the expression of beauty at the sight of the instrument, are found only in some men. It follows, then, that when we see the expression of the sense of beauty in a man at the sight of the instrument, this indicates one of his properties, one of his forms of impressionability. In fact, the expression of the sense of beauty at the sight of a certain object is a reflex movement of a certain kind. I regret to be unable to describe the expression of the sense of beauty in man, but I think the following are important points in such expression :—The sight of the object causes the head and eyes to turn towards it, as when the attention is attracted; then the visual impress of the object is usually inhibitory, rather than stimulating; movements are commonly inhibited thereby, and the primary appearances of rest are produced, just as when the attention is attracted by anything; sometimes, as in children, there may be signs of excitation rather than inhibition. There is, I think, sometimes a tendency to extension of the head, which may even occur to such an extent as to remove the eyes from a position where they can see the object. These motor signs at the sight of a particular object occur only in some people; they serve as physical signs indicating what kind of brain they have, or, as it might commonly be said, what knowledge and power of appreciation they have. A well-dissected anatomical preparation may excite in one man the expression of the sense of beauty, but in another it may excite the expression of mental pain. The series of reflex movements produced in the two observers by the sight of the preparation are different. The difference expresses certain properties of brain in each man.

CHAPTER XIV.

EXPRESSION OF MIND IN THE INFANT AND ADULT.

Materialistic questions only entertained—The criteria of mind, what are they?—Physical study of signs of mind from infancy upwards—A subjective condition is only known to us by its physical expression—Brain properties necessary to mentation—Impressionability—Retentiveness—Relation of outcomings to afferent stimulus—Comparison of an infant with the adult, and an idiot with a healthy child—Description of an infant; its development and signs of potentiality—Impressionability; its attributes; delayed expression of impressions—Modes of expression are criteria of mind—Expression of distress—Memory—Subjective conditions studied by their expression—Thought.

THE physiologist and the physician, dealing in their work only with things material, must of necessity seek for realistic methods of investigation and description. We have not here to consider any metaphysical properties of mind, but, putting aside all such considerations, it is our business at present to consider only the realistic, objective, physical signs by which the physical investigator may judge of the presence of what is called mind.

What are to be taken as the criteria of mind or the faculty mentation? We leave this question to be partially answered in the course of this chapter.

It has been very usual in inquiries into the faculties of mind to refer to the structure and properties of the adult brain, and then to argue from that basis all round the subject. I propose here to consider first the human infant, seeking for physical signs of its mind, or of the faculties of mind actual or potential, and to trace the signs of its mental development upwards towards adult age. The object of my work is to give such description and definition of the conditions of the brain when exercising the faculties of mentation, as may enable us, as observers and experimenters, to note with exactness how certain forces acting upon the man elicit the signs of mentation, and generally to attempt a physical experimental inquiry into the function of mentation. It has often been said by authors that we study mind by recording subjective feelings, their associations, and combinations, but we can only know the subjective feeling of another man by its physical expression in movements, speech, and other results of movements.

Our method of inquiry is based upon observation and experiment, and is intended to elicit knowledge as to the action of external forces as factors in the evolution of the faculty "mentation." On account of the nature of this inquiry and its objects, it is necessary to deal with the question purely from a physical point of view, so that we have nothing to do with "feelings" or "consciousness" as elements of mind, for they cannot be directly observed or experimented upon. It is also necessary in each portion of such an inquiry, before giving observations and experiments as proof of points concerning any property or function, to decide upon physical signs which are to be considered as criteria of that function. It is only these physical signs that we can observe or experiment upon.

The principal properties of the brain necessary to mentation are—(1) impressionability, the effect and outcome of the impression varying according to the afferent force or stimulus; (2) retentiveness or permanent impressionability; (3) special association of the outcomings as dependent upon the afferent stimulus. Or we may say that the expression of these properties in part indicates the potentiality for mentation. One method of determining the physical signs of mind is to compare subjects possessed of mind with others devoid of mind, or nearly so.

It will be granted that an infant at birth does not show well-marked signs of mind; the principal signs of mind are absent at birth. An adult man who presented no more signs of mind than a child at birth, would be said to be "mindless"—in a condition of amentia. An infant at birth may be said to possess none of the actual faculties of mind although it is healthy; it may possess potentialities, but it shows no actual, present signs of mentation. An idiot in growing up from infancy, does not show those objective signs appropriate to its age which indicate potentiality for the functions of mentation. To catalogue the signs which indicate idiocy, is to summarize the signs of the absence of mind (see chap. ii. p. 13). The infant is said not to show actual signs of mind, though it may show potentialities. The infant does not walk, talk, or turn its eyes and head towards a bright object within its field of vision; its movements are not modified in any marked degree by the action of light or sound, except that the orbiculares oculi contract spasmodically to light. The infant is less impressionable, and the impressions produced by light, etc., are less permanent than in the adult.

Again, to speak of infant development, the signs of its brain development are identical with the signs of its mental development. Speaking of either mode of development in an infant we may give its expression.

The healthy infant born at full time weighs between six and ten pounds; its limbs and members are complete in all parts—fingers, toes, nails, etc. The head measures in circumference from eleven to twelve inches; the sutures or junctions of the bones of the vault of the skull are not closed or ossified, and the anterior fontanelle is open. We may also observe the form, size, and proportions of the body, and particularly of the head, as signs which indicate to some extent the degree and condition of brain development.

Respiratory movements in the infant are established at birth, and continue without interruption; the child cries when its skin is cold or wet, and when the stomach has been empty more than two hours.

The nerve-muscular mechanism and the sensory

surfaces allow of the occurrence of certain reflexes. An object placed in the mouth stimulates the movements of sucking; cold to the skin is followed by crying; light causes contraction of the orbicular muscles of the eyelids, and if the eyelids are raised the iris contracts to light. The tone of the sphincter apparatus enables the hollow viscera to retain their contents. In a newborn child a few hours old, the attempt to straighten the elbow when flexed may be strongly resisted.

Frequent spontaneous movements may be seen while the infant is awake; movements apparently irregular, are almost constant in the hands, fingers, and toes. A short period of wakefulness is usually followed by sleep, indicated by subsidence of movement in the limbs and closure of the eyelids. We say that the newborn infant does not give expression of the faculties of mind, because it does not present physical signs showing that it is impressed, even temporarily, by the sight of surrounding objects; it does not move its hands towards objects within its field of vision, and no movements indicate that it is impressed thereby—reflexes of sight and sound are almost entirely absent.

In the early stages of development there is inability to put out the hand, moving it in a straight line towards an object. There is also inability to touch a definite part of the body where there may be pain. When one leg itches the child is unable to scratch it with the hand, but tends to do so with the other foot. The power to grasp an object by its own act is a later develop-

Fig. 71.—Tracings of the spontaneous movements of an infant's hand during fifteen minutes.

ment. The ability to transfer an object from one hand to the other is not acquired for some months, and the association of the two hands in playing with an object is still later in development.

Very important signs are the movements indicating recollection upon hearing the names of well-known objects (retentiveness). In such a case only waves of sound act upon the brain from without, and the movement that results in the child, indicates that the sound has been heard. A certain name may always cause the same facial gesture (expression); thus, "bottle," "bed," excite the facial expression of pleasure, disappointment. The word is said, and the face changes. However complex in mechanism and function the brain which moves the face may be, still it is the sound that stimulates the movement.

Retentiveness of the effects of the sight of an object is very important. Does the child remember objects shown to it? Here the "memory" is indicated by the movements the infant makes when shown an object that it has had sight of often before. The undulations of light are reflected from the particular object; these, falling upon the retina, produce, by their action on the brain, the gestures in the face, hands, etc., the expression of joy, pain, etc.; the sight of this special object always producing similar effects, subject to modifications. The sight of a funny doll makes the child laugh; a dog makes it cry and clench its fists.

Upturning of the eyes in their orbits, accompanied by elevation of the eyelids, in looking at

an object high up, is a late development in the infant, and indicates the advance of intellectuality.*
It is noteworthy that in hydrocephalus the eyes frequently roll upwards, but the eyelids remain drooped, hiding the corneæ.

In the list given above, which is by no means complete, the order followed is about that in which the development of these signs takes place. We will now attempt some kind of physiological classification. Some of these nerve-muscular signs follow and result from external agencies. Thus, turning towards a light or an object depends upon light passing from these things to the child's eyes and brain;† turning to a sound is the direct effect of the sound-undulations of the air. So, again, when movement is checked or altered by "the sight of an object" or "a sound," the result is the effect of the waves of light, or of air upon the nerve-mechanism; in every healthy brain the effect of such waves is probably nearly identical. Other motor signs are more directly intrinsic, the physiological outcome of the structure and properties of the nerve-mechanism, thus: crying, reflex actions, parallelism in the movements of the eyes, the constant movement of the fingers and limbs while the infant is awake, some movements of expression in the face, playfulness; and later on, the repetition of identical movements upon similar stimulation, the acquired power of co-ordinating movements of the hand; and still later, the power of using both

---

* Chap xii p 223.   † See chap. x. p. 185.

hands together—these are the outcome of the properties of the nerve-mechanism itself.

The following observation on one of my children when eighteen months old illustrates how the dawning intellectuality is indicated by the complication and fitness of certain sets of movement. "The child having both hands full of toys, desired to grasp a third; he then put the toy from one hand quickly between his knees, and thus set one hand free to take hold of the desired object."

The following kinds of movements as signs of a healthy infant brain deserve separate attention:— Movements following certain external agencies, light, sound. Movements the outcome of the essential (untrained) properties of the nerve-mechanism. Movements resulting from the acquired association of nerve-centres (training). Movements similar to those previously occurring from a like cause, showing retentiveness. Movements in different areas, such as the small joints in contrast with large joints; or a different condition of movement of adjacent parts, such as the fingers. There may also be asymmetry of movements.

The muscles of the face are seen to act earliest in the lower zone,* those about the mouth causing expression before those on the forehead (corrugators), which seem to be specially connected with mentation. If the organization of the infant is not very strong, the eyes as they move in their orbits do not maintain a strict parallelism of their axes;† this is

* Compare with cerebral facial palsy, see p. 108, chap. vii.
† See account of eyes in sleep, chap. xii. p. 218.

most commonly observable while the child is sucking from its bottle.

Now, as to the *child when four months old*, we say that the attention is easily attracted, because the sight of objects and sounds cause the head to be moved (by reflex action) towards the light or source of sound. More than this, after the stimulus of the sight of an object has caused the head and eyes to be turned towards the object, the further stimulation of the brain may inhibit the kinetic functions of the brain, arresting all movement;* this often happens when the attention is attracted. On the other hand, the sight of an object, after it has caused the head and eyes to be turned towards the object, may increase the amount of movement in the child. The difference in the effects of the visual impression is an expression of the condition of the child.

Playfulness is probably the result of spontaneous movements, together with an increased susceptibility to reflex action. The "playful child" has a happy face, owing to the healthy tone of the facial muscles and their nerve-centres.

The table given on page 250, is for a healthy, well-developed infant of good class.

Perhaps of all the means of expression enumerated, those which indicate impressionability are the most important, and in some particulars the most difficult to understand.

Thoroughly to grasp this conception we must consider the attributes of the property " impression-

* See chap. vi. p. 101, Fig. 9.

ability"—quantity, time, kind. The impression made at one time may not be immediately followed by an outcome; the outcome may be long delayed, a mere question of the attribute time; it may be delayed till certain forces again act upon the subject An impression may be latent till certain circumstances call the effect of that impression into activity.

Age in Months.	Weight.	Head Circumference.	Point indicating Stage and Progress of Development.
	lbs.	inches.	
1	7·10	14·5	Power to suck; regular succession of feeding and sleeping; hand reflex.
2	11·0	15·25	Hair in eyelashes and eyebrows; may be occasional strabismus.
3	13·5	16·5	Capability of shedding tears; no strabismus.
4	15·0	17·0	Constant movement while awake.
5	15·5	17·0	Turning head to a light or sound.
6	16·0	17·25	Recognizing objects, as mother, nurse.
7	17·5	17·5	Holding object in hand, and carrying it to mouth.
8	18·5	17·75	Various sounds made; commencing dentition.
9	19·5	18·0	Some power to hold up head while lying down.
10	19·6	18·25	Can hold an object without dropping it.
11	19·7	18·4	Power to transfer object from one hand to the other.
12	20·0	18·5	Commencing to crawl or stand with assistance.

We cannot know the impression upon the brain itself; we only know the effect in the outcome of that impression.

A child of four years old quietly looks at some one putting a letter into a pillar post; we cannot at the time see the impression produced upon the child's brain, but we guess that an impression has been produced because the child's head and eyes turned towards the pillar post. We know that an impression has been made when next day, on the child finding a letter on the table, "he takes it and posts it behind the door."

The attribute time, of the property impressionability, is very important as a factor in this case; the series of the child's movements on the second day, under the stimulus of the sight of a letter, are the expression of its impressionability and of the more or less permanent nature of the impression, and as such are evidence of the faculty mentation. An idiot would not do so. It is, then, very important to appreciate the attribute time of the property impressionability. In the above example the kind of impression is susceptibility to the production of a series of movements upon the visual stimulus of a letter; the quantity, or, as it is often termed, the depth of the impression is indicated by its durability and the readiness with which the special series of movements is excited.

One of the great signs that an infant possesses potentially the faculty mentation is, that its attention is easily attracted to the sight of objects, and to a light. Such stimulus produces expression by movement at an early age in a healthy infant. The head rotates, and the eyes turn towards the object or the light, and spontaneous movement is stimulated or inhibited. It is well known that a moving object especially attracts the attention, if the movement be not too rapid. Can we give any explanation? As the object moves, different portions of the retina will be stimulated; and the greater the area of retina stimulated, the greater the stimulus transmitted to the nerve-centres which move the head and eyes, the greater also the inhibitory or stimulating effect produced upon the spontaneous movements.

In the *adult* the objective *criteria of mind* are *modes* of *expression;* the expressions of the emotions, feelings, passions, thoughts are indications of the mind; and all these modes of expression have been shown to be produced by direct action of the nerve-system. It is, then, admitted that conditions of the mind are directly expressed by nerve-muscular signs. This implies that some material, physical change occurs along with "mentation," which material change is expressed in the muscles of the body. It is this inherent physical change, thus directly expressed, which the physiologist investigates in his studies of mind; and whether that inherent nerve-change, thus directly expressed, be mind itself, or in some way allied to mind, is a metaphysical question I shall not attempt to discuss. Here we only discuss material, physical action.

Let us take an example or two. A mother just after the loss of her infant comes to speak of its death. We then see the face tending to flush, the mouth quivering, the angles depressed; she speaks with a trembling voice, half choked by her emotion.

The term "mind" is used as the collection of all the properties that make up the faculty mentation. We cannot by physiological methods investigate "mind" itself, as we know not what it is, but if we agree to consider certain physical signs as indices or criteria of mentation, these criteria can be observed. Thus *memory* is an abstract phenomenon, but by processes of analysis we can determine certain criteria as indices of the property "memory," and these physical signs can be observed and even experimented upon, thus enabling us to deal by physical processes of inquiry with the faculty "memory."

Subjective conditions, pain, joy, fear, etc., can be studied by observing their modes of expression. It is the expression that we can record and analyze. When a child touches a hot teapot his movements and combinations and sequences of movements convince us that he felt the heat. An idiot, when he touches the hot teapot, does not move much, may not move at all; this would be taken as a sign of his mental deficiency. A very strong-minded man of the Spartan type might inhibit the reflex movement stimulated by touch of the hot object. An object impressing by sight, such as a beautiful flower, may inhibit the spontaneous movements of an intelligent man. It will not so affect a blind man, and it will produce no effect, present or prospective, on a comatose man; he is not at all impressionable. Similarly, a deaf man is not stimulated to movements or otherwise impressed by sounds. Impressionability is one of the expressions of mentation.

No account of the methods of studying mind can be complete without reference to the processes of thought. Now, "thought" is not a physical thing, and is not therefore within the range of our consideration, but we can deal with the expression of thought. We know that a man has thought by what he subsequently says and does. While deeply thinking there is but little movement, but little outward expression; the kinetic functions of the brain are suspended while it is performing its mental functions. The subsequent series of actions, the sequential modification of spontaneous movements, and the susceptibility to new reflexes and fresh impressions—these are the expression of the unknowable process thought or mentation.

Physical study of mind. Table of brain properties necessary factors of the faculty of mentation. It is interesting to notice that the properties of the brain which indicate its potentiality for mentation, are analogous to properties found in plants and inorganic things:—

Impressionability: phonograph, telephone (pp. 18, 19).

Retentiveness: phonograph (p. 19).

Reflex action: Drosera (p. 152).

A series of actions: (*a*) resulting from an apparent force,—paper organ machine; folding of leaf of *Mimosa;* (*b*) a combination or aggregation of actions as the result of a stimulus.

Change of function: see p. 39.

Inhibition: see p. 94.

Nutrition: see p. 234.

CHAPTER XV.

ANALYSIS OF EXPRESSION.

Analysis of the expression of fatigue—Localize the expression—Observe trophic signs, postures, movements—Analyze and classify movements according to the principles given—Analysis of Darwin's description of laughter, and Sir Charles Bell's description of joy—The importance of such analysis—Pope's description of Achilles—Study of a nervous subject—" A school inspection "—National modes of expression.

LOOKING at a certain man we recognize that he is fatigued. How can we analyze the expression of fatigue? Being sure upon general knowledge that the man is fatigued, we seek to analyze the expression of fatigue. The first process is observation. We may commence by trying to localize the signs of fatigue in certain parts; look at the head, face, trunk, limbs, their postures and movements; also look to the trophic signs, the form and fullness of the face, etc. In all these points compare what is seen with what experience teaches would be seen under like circumstances if the man were not fatigued.

As an example, a man is seen sitting in an armchair, doing nothing but talking. The limbs are not free; the feet rest on a foot-stool, the hands rest on the sides of the chair, the head rests on the back of the chair; the face is free and the eyes and tongue are also free (their nerve-centres are free) to be acted on by external stimuli. It is, then, to the face, eyes, tongue that we must look for the best signs of expression in the man. We may then request him to hold out his hands, that we may observe the spontaneous postures that follow. We study his movements, and results of movements, analyzing the face, observing the movements of the eyes, and noting the speech, at the same time describing the postures seen; by so doing, all the motor factors in the expression may be analyzed and classified according to the principles of analysis and classification of movements given in chap. v.

We want to lay down some kind of rules for the analysis of modes of expression; thus—

1. Localize the expression.

2. Notice whether the kind of expression be trophic or motor; if motor, analyze it according to the principles * for analysis of movements, giving an anatomical description.

3. If there is movement, is it primary and spontaneous? or is it stimulated by external † forces, and thus to be considered a reflex action?

4. An element in an expression may be the inhibition of spontaneous ‡ movements usually present.

5. The mode of expression may be trophic in

* See chap. v. p. 75. † See p. 100. ‡ See pp. 58, 64.

kind; if this be so, search to see if it be a mode of expression direct, or by coincident development.*

If my principles of expression and descriptions of the modes of expression be somewhat true to nature and practically useful, they will enable us to analyze any good classical description of expression in man, if given in physical terms. Taking any such description, we ought to be able to analyze it, and replace, if necessary to our purposes, the terms used by the author by our own terms, thus giving his account in such terms as enable us to submit the description to the test of direct observation. Such points in our author's description as cannot be thus translated suggest that either our principles are defective and incomplete, or that the author's description is metaphysical or imperfect from our point of view.

If we take C. Darwin's † description of laughter, we find it given in terms of nerve-muscular signs, as are almost all his descriptions:—

"During laughter the mouth is open more or less widely, with the corners drawn much backwards, as well as a little upwards; and the upper lip is somewhat raised. The drawing back of the corners is best seen in moderate laughter, and especially in a broad smile—the latter epithet showing how the mouth is widened. . . . Dr. Duchenne ‡ repeatedly insists that, under the emotion of joy, the mouth is acted on exclusively by the great zygomatic muscles, which serve to draw the corners backwards

---

* See p. 273.   † "Expression of the Emotions," p. 202.
‡ "Mécanisme de la Physionomie Humaine," Album, Légende vi.

and upwards; but, judging from the manner in which the upper teeth are always exposed during laughter and broad smiling, as well as from my own sensations, I cannot doubt that some of the muscles running to the upper lip are likewise brought into moderate action. The upper and lower orbicular muscles of the eyes are at the same time more or less contracted; and there is an intimate connection, as explained in the chapter on weeping, between the orbiculars, especially the lower ones, and some of the muscles running to the upper lip."

Sir Charles Bell * says—

"In joy the eyebrow is raised moderately, but without any angularity; the forehead is smooth, the eye full, lively, and sparkling; the nostril is moderately inflated, and a smile is on the lips. In all the exhilarating emotions, the eyebrow, the eyelids, the nostril, and the angle of the mouth are raised. In the depressing passions it is the reverse. For example, in discontent the brow is clouded, the nose peculiarly arched, and the angle of the mouth drawn down."

Having these descriptions before us, we can make some comparisons, or analogies, and can apply the principles of antithesis. In laughter, which is an expression of joy or happiness, we have material problems to deal with capable of physical investigation. The angles of the mouth are said to be drawn upwards; this is the very opposite to the expression of physical suffering. By defining the

* *Op. cit.*, p. 172.

expression of the abstract thing, happiness, in terms of nerve-muscular signs, we find material problems to deal with capable of physical investigation. John Bulwer's* descriptions of expression were given in terms of nerve-muscular signs.

In Bell's account of joy, the first paragraph is in terms of nerve-muscular movements; then comes the paragraph, "the eye full, lively, and sparkling." This is an artistic expression I fail to analyze. Does the "full eye" mean a condition of parts outside the eye? Does the term "lively" apply to a muscular condition? Does the "sparkling" eye depend on tension due to muscular action? I trust that further knowledge may enable us to include these under our principles of expression.

Analyzing Bell's description, we find, then, that with the exception of one paragraph the terms used are all movements and results of movements.

What good, what advantage is there in these special modes of describing what we see? Our modes of description are such as allow of comparisons being made. We translate abstract quantities such as "joy" into concrete terms, such as nerve-muscular signs, or conditions of form or development. We translate the terms used to describe the abstract property into other terms the expression of the abstract. The term "happiness" is a word intended to indicate a certain condition of feeling which we all more or less understand; the thing happiness in the abstract is an abstraction, it has no material existence as an entity; but if we can

* See chap. xviii., p. 323.

define an expression of happiness in man, we can deal with the material expression of happiness, analyze it, study the coincidences and successions of movements seen in this condition and their concomitants, etc.

Sometimes a term is used in common language which cannot be thus directly translated, but requires analysis, such as the common term "jolly;" thus, "John looks jolly." There is no metaphysical expression "John;" and the modes in which his condition is expressed are material: analysis is then possible. If we observe a lot of jolly boys, I think we shall agree that the predominant characters are good nutrition and active movements. "Jolly" is, then, a compound condition, partly trophic and partly kinetic; the movements, or kinetic brain function, can, of course, be analyzed and its components classified as in any other expression by movements.

We will now take for analysis Pope's description of the condition of Achilles:*

> "Achilles heard, with grief and rage oppress'd;
> His heart swell'd high, and labour'd in his breast.
> Distracting thoughts by turns his bosom rul'd,
> Now fir'd by wrath, and now by reason cool'd."

This description we should find hard to analyze according to our principles, on account of the number of metaphorical expressions used.

Instead of analyzing an author's description let us now consider a common example—"A lady seen in her drawing-room looks nervous." This may be

* Pope's translation of the "Iliad," line 251.

one's general, social impression. How shall we proceed by scientific process to analyze the expression before us. Observe the permanent conditions of development—the head, the build, and proportions of the body. As to the head, note its size, proportions, form, and angles; the shape of the facial features—ears, lips, eyelids, nose; note whether the form on either side be symmetrical. Observe the postures seen on the average, or occasionally, especially the postures of head, head and eyes, spine, hands, and face; note signs of symmetry and coincident movements. In looking for movements note also whether the hands and other parts are kept free, or whether they are purposely engaged, as in holding book or fan,—this mechanical occupation of the hand preventing the fingers from twitching and performing meaningless antics, and so preventing the nervousness and fidgetiness from being socially observable. In this nervous lady the hands will probably be kept not free,—she will, unless very young, have learnt to keep her hands engaged when talking, that she may not gesticulate unduly—but when her hands are free, finger-twitching may be seen. When her hand is held out, or if it hang free over the arm of a chair, we may see "the nervous posture" or "the feeble hand;" if this be seen, look to the posture of the other hand, and note the coincident posture of the head, which usually lolls over away from the side which shows most markedly the nervous hand. Among head postures and movements, slight flexion with rotation and inclination to the same side is common in the nervous condition;* there may be occasional extension or lifting up of the head so as to carry the eyes upwards and place their field of vision above objects around.†

Among the muscles of the face there is likely to be over-action, especially of the occipito-frontales and zygomas, causing respectively "the surprise look" and "very broad smiling;" this latter is often unequal on the two sides. These and other points may indicate, after analyzing the lady's expression, that she is nervous. The erect head without inclination shows general firmness; frequent inclination with rotation to the same side expresses a certain amount of independent action of nerve-centres, such as is seen to excess in nervous people.

Fidgetiness of fingers shows a great amount of separate action of small nerve-centres, or the centres for small parts; thus by analysis, and applying the principles for analysis of movements, we may study the lady's nervousness by analyzing its expression; at the same time it may be noted what circumstances increase and what diminish it.

At a school inspection the children may be observed, and the expression of their condition may be analyzed. In carrying out such a work it would be well to make observations under different circumstances: (1) while the children are at their ordinary school work; (2) while under the inhibitory influence of being told to keep silence; (3) when at play; (4) a personal inspection may be made of the individual child.

* See chap. x. p. 188.   † See chap. xiii p. 238.

Now as to what to observe, and how to analyze the expression seen at the same time.

I think that in practice the most important observations are most easily to be obtained by procedure in the above order; but, for convenience of analysis, I think it convenient to consider the last point first. On looking at an individual child there must always be some expression; it may be expression of the normal, the quiescent condition. The trophic signs may be normal in due proportion, and the nerve-muscular conditions may be in rest. In making observations we look for the following points:—

(*a*) Spontaneous movements, and postures.
(*b*) Reflex movements.
(*c*) Trophic conditions.

There may be no movements seen, or but very few. Absence of spontaneous movements is the absence of signs that ought to be present as being signs of health. It may be, however, that the brain is healthy though it produce no movements; the motor power may be inhibited for the time being. This absence of movement may be due to a want of nutrition of the nerve-centres; the brain may be so exhausted as to be unable to produce the normal quantity of movement.* It follows, then, that an important question arises. Is the absence of movement due to some external force preventing the display of the kinetic action of the brain? or is it due to exhaustion of the child, resulting from ill nutrition? Now, then, we may look for other signs

* See description of fatigue, chap. xiii. p. 228.

indicating the condition of nutrition * of the body at large, of which the brain is a part; we note the present size and proportions of the body, the fullness and fatness of the face, its paleness or colour. In looking for causes of ill nutrition of body and brain, we should, of course, examine for the signs of disease,—consumption and heart-disease may cause ill nutrition of body and brain; but, as this is not a medical essay, we pass over such matters here.

While looking to the general trophic conditions, the observer should look out for the signs of any special diathesis. In chap. xvi. a description of the signs of the scrofulous diathesis is given; to detail the signs, which indicate certain physical tendencies, is important in analyzing a child's condition.

If there be signs of good nutrition, with the absence of movements, we may suspect that the kinetic function is inhibited. Inhibition of spontaneous movements upon due occasion is a very important result of school training and discipline. A certain degree of inhibition of movements is necessary to the display of intellectuality; it is a valuable result of training when the self-contained power of the child stops its movements, and makes it sit or stand still. It is obviously a very different matter to find, as the result of analysis, that a child's movements are inhibited at will, to what it is to find that they are absent from ill nutrition.

To examine the child's nerve-muscular system

* See expression of nutrition, p. 230.

further and analyze the results, let him hold out his hands free, then some of the spontaneous postures described as expressive of certain brain conditions may be seen, such as the straight hand, the feeble or the nervous hand. The condition of the facial muscles, and in particular the orbicular muscle of the eye, should be specially noted.* Finger-twitching may be seen at the same time.

Inasmuch as many nerve-muscular signs in the child are reflex actions, we should vary the circumstances surrounding the child: we speak to him, show him things, and see how he is impressed or affected by them; that is, we observe the outcome following certain afferent stimuli. If there be any doubt about the condition of the special senses, we note if his head and eyes turn towards a sound or a light.

We note, then, signs in movement, in nutrition, and in proportional development, and consider what they respectively express.

In class we may observe the children's movements of various kinds:—

(α) Spontaneous movements.

(β) Spontaneous movements modified by surroundings.

(γ) Spontaneous movements inhibited in part.

(δ) Reflex movements.

Observing the children at play enables us to form some kind of judgment as to the amount of spontaneous movement; and by comparing the action of a child at play with its action in class

---

* See chap. xi. p. 206.

we form an idea of its susceptibility to inhibition of movements.

It would be an interesting and curious inquiry to analyze and describe the national modes of expression and gesticulation. I was much interested one day to meet in a railway carriage an Englishman, an Irishman, an Italian, and a Frenchman, engaged in animated conversation; the gesticulations of each were different, and characteristic of the respective countries of their birth.

CHAPTER XVI.

CONSIDERATIONS AS TO THE ATTRIBUTES OF A PROPERTY OR FUNCTION—TIME, QUANTITY, KIND; AND AS TO THEIR RELATIONS.

The attributes of a property or function: time, quantity, kind—Attributes of the functions trophic and kinetic action, in one subject, in two or more subjects—Combinations and sequences of action—Proportional growth; equal proportional growth; similar development—Analogy between series of kinetic and trophic actions — Special combinations of action may result from afferent forces; this an important element in evolution—Heredity—Summary.

ANY property or function may be described as having the three attributes—(1) time; (2) quantity; (3) kind.

As examples of vital function, let us consider growth, or trophic action, and movement, or kinetic action. Concerning every function or property, whether it be physical or vital, we may describe the attributes time, quantity, kind.

As to *Time;* we may observe the moment of action, frequency, duration of the active appearance, and of the quiescence or disappearance of the function. Time is the only attribute of totally dissimilar functions that can be directly compared.

As to *Quantity;* if we can observe and measure this at each period of time, we thus determine the total quantity of the function. To describe the quantity of a function, it is necessary to adopt a unit of quantity. If any unit of quantity can be found as common to two dissimilar functions, a comparison may be made between them as to quantity.

As to *Kind;* it is not convenient to consider this here. In some cases the "kind" may be described in terms of time, and quantity of the function or action. Nerve-muscular movement has practically only two attributes—time and quantity. What we commonly call "kinds of movement," are usually series of movements, each separate movement having its own time and quantity. It may be here remarked that one of the great differences between the properties of living and non-living things is, that in the former properties often vary in time, quantity, and kind; whereas in non-living things a property, such as hardness, is more permanent in its duration, and more unchangeable, except under the agency of obvious afferent forces. Iron retains its hardness except under the influence of heat.

When we come to consider the attributes of growth, or trophic action, and movement, or kinetic action, in two or more subjects, other points present themselves to our notice.

As to the attribute *Time* of growth or movement in two subjects. The time of growth or movement in each subject may coincide, or it may not coincide.

With three subjects, A, B, C, growth or movement may occur in each separately, or we may have coincidence in the following combinations: A, B, C, AB, AC, BC, ABC. We may, then, in considering the growth, or any other action or function, in two or more subjects have coincidences, combinations, sequences, series.

As to the attribute *Quantity* of growth or movement in two subjects. We may find equal quantities (in equal times) in each subject, or unequal quantities. We may, then, in considering the growth, or any other action or function, in two or more subjects, compare the ratios of growth. We may, in two or more subjects, compare the same function as to the attributes time and quantity.

I. "Arthur and John are twin brothers, and they are very much alike."

Further observation and analysis may enable us to write thus:—

II. "Arthur and John are twin brothers; they are alike in height, form, and proportions of the head and body; they are also alike in their manner of walking and speaking."

Paragraph I. is a mere arbitrary statement, presenting a proposition. Paragraph II. is a proposition giving also some evidence. Let us analyze the evidence, and the statement made.

(1) "They are alike in height, form, and proportions of the head and body." All these points are statements concerning trophic action, and they are capable of further analysis and description.

(2) "And they are alike in their manner of walk-

ing, and speaking." These latter statements concern kinetic action.

Further to analyze the statements, (1) here we have examples of equal proportional development; (2) here we have examples of similarity in walking and speaking. Both walking and speaking consist of a series of movements. If Arthur and John have a similar manner of walking, that can be described and demonstrated in terms of a series of movements; it is a series of kinetic actions that is said to be alike in Arthur and John.

It remains to explain the terms "equal proportional growth" and "a similar series of kinetic actions." When we speak of proportional growth we really refer to two subjects, and we compare in each the function growth as to its quantity.

We will first inquire what is meant by "proportional growth," and then consider some of its variations.

Proportional growth.—This term, as it concerns a proportion, must imply something about the quantity of the growth; it is only as to quantities that we can make a proportion or ratio. It concerns, at least, two subjects, and implies the condition that the ratio of growth of two or more parts or members of one subject, is equal to the ratio of growth of the corresponding parts or members of the other subject.

When we speak of proportional growth in John, we really consider two subjects, not one; we are using in reality a condensed phrase. Statement (1) speaks of equal ratios of the head and body in

Arthur and John; the ratio of the quantity of growth, and the size which indicates the result of growth, in the head as compared with the body is identical in each man.

When we speak of equal or good proportion, or otherwise qualify the kind of proportion, we are in reality considering at least four things. Thus, when I say that a child's head is in good proportion, I compare the size of its head with the size of its body, and then compare this ratio with the ratio of the head and body in a perfect child.

Equal proportional growth.—This term is convenient, but it requires explanation. The meaning of "proportional growth" has been explained. It was found that the term was applicable to two subjects: it concerns the ratio of growth, and denotes the proportion of the quantity of the growth in the one subject to the quantity of growth in the other subject.

The term "equal proportional growth" implies that two ratios are equal. Two subjects, each presenting two similar parts for comparison, are here referred to; the ratio of the quantity of growth in the two parts of the one subject is said to be equal to the ratio of the quantity of growth in the two corresponding parts of the other subject.

"In Arthur and John the proportional growth of the head and body were the same."

The quantity of growth in the head and body of each man was observed; it then appeared that the ratio of the quantity of growth of head and body in each was the same.

Good proportional growth.—We often say that a certain man is well proportioned, or that he presents good proportional growth. This is one mode of expression of beauty in the human form. In speaking of a proportion we must refer to two subjects, as to the quantity of growth. When we say that the proportion of growth in two parts of the subject (head and body) is good, we are really making a further comparison. We compare the ratio of growth observed, say, in John, as to his head and body; and we compare this with the ratio found on inquiry between corresponding members in a perfectly formed model: the "goodness" of the proportion depends upon how near the ratio of the size (quantity) of head and body in John approaches to that of the perfect model. A good proportional growth depends upon the comparison of the ratio actually observed with the similar ratio of the perfect model. Thus, when I say that a child's head is in good proportion, I compare the size of its head with the size of its body and thus obtain a ratio; then, ascertaining the ratio of head and body in a perfect child, I compare the two ratios, and contrast the actual with the perfect proportional development. If John's proportional growth is the same as that of the perfect model, we say that John is well proportioned. It seems, then, that the "kind" of development may be a question of proportion.

The consideration of the term "a similar series of kinetic actions" is postponed (see p. 276).

In examining the attributes of functions, we have

inquired as to some points concerning time and quantity. Now, as to the attribute Kind, I have no suggestions to give here as to the essential character of kinds of growth. Observation shows that the kind of growth occurring in a part of the body may be good, bad, or indifferent. It generally happens that if the growth of parts of the body that can be seen is good, the growth of organs that cannot be seen is likewise good. This is an empirical statement.

This leads us to study a mode of expression referred to in chap. iii. p. 42, under the term "Coincident similar development."

It often happens that after extended observation we find the condition of growth or development of one part of the body expresses, or indicates, the presence of certain properties in the subject not directly connected with the special sign observed. Dull heavy-looking features usually accompany a dull inactive mind (brain), not because the dull features cause an inactive condition of mentation, but because this cast of features usually accompanies a make of brain with slow action. This mode of expression is, then, indirect and empirical; in many examples no causal connection can be seen between the two subjects of similar growth. In an Englishman excessive development of the epicanthic fold of the eyelids (chap. vii. p. 137) is often accompanied by mental dulness; here, then, there is similar bad development in eyelids and in brain. The want of symmetry of bilateral or corresponding parts often indicates poor development of brain. Asymmetry of ears in dull children is common.

Similarity of development in two parts may be good in kind. Handsome, regular features of the face often accompany mental perfection. In giving a list of examples of similar development it will be seen that some are empirical, and not at present capable of explanation; such are probably due to a common force acting alike on both subjects. In other cases the coincidence can be explained.

Emaciation or fatness of the face usually (not always) indicates emaciation or fatness of other regions of the body.

Absence of the organ of hearing indicates deafness, and usually indicates deaf-mutism.

Congenital absence of eyes indicates blindness, and probable atrophy of the optic tracts.

A congenital condition of the skin, termed icthyosis, indicates usually a liability to bronchitis; in this condition the skin cannot perspire, and too much work is thrown upon the lungs.

Plants with small ill-developed leaves show much constitutional delicacy; the small leaves cannot assimilate material enough for nutrition.

A very small or microcephalic head is a sign of congenital idiocy. The absence of the faculty mentation depends upon the smallness of the brain consequent upon the microcephalism. If the skull of the infant at birth be fully ossified expansion will not occur, and the brain must ever remain microcephalic; but still it is not certain that the brains of such children would be capable of growth and development if the bony case were expansive as in a normal child.

Cleft-palate may accompany marasmus; both may be due to similar ill development, or the cleft-palate may cause difficulties in feeding the child and lead to marasmus. Congenital collapse of lung may lead to patent foramen ovale in the heart.

The principle has long been admitted that the tendencies in the development of a child or adult may be studied by determining the diathesis, as it is called. Certain things are observed in the man, and then experience enables us to say that such and such will be his constitutional tendencies.

When we observe in the individual that certain functions are of one kind, we infer therefrom the kind of other functions. Thus, Mr. F. Treves[*] gives the following description of the strumous diathesis, and shows such persons liable to certain pathological conditions: "The physiognomy of scrofula, the type of face and form supposed to be indicative of the disease, have for ages been subjects upon which writers have loved to exercise their imaginative and descriptive powers" (p. 83). "The general features of this class are sufficiently well marked to enable us to separate them into two divisions, that, for want of better words, may be known by the old terms—the *sanguine* and the *phlegmatic* types of scrofula" (p. 85).

"The *sanguine type*. Individuals placed in this class are credited with these features, and they refer more particularly to children. They are tall,

[*] "Scrofula and Gland Disease," 1883.

slight, and graceful, with well-formed limbs, hands, and feet, a fine clear skin, and usually a fair complexion. The face is oval, the lower jaw small, the features delicate and regular, the lips thin. The eyes are bright, and covered with long eyelashes, and the hair is often remarkably fine and silken. A sprightly and excitable disposition may be added, and the picture is complete " (p. 86).

"In the *phlegmatic type* are comprised individuals, as a rule, short and burly, with coarse limbs, large hands and feet. The face is broad, the lower jaw large, the malar bones often prominent, the features coarse and irregular. The nose is generally thick, the lips tumid, the lobes of the ears large, and the neck unshapely. The skin is coarse, harsh, and thick. The amount of subcutaneous cellular tissue is considerable, and often sufficient to conceal the muscular outlines of the body. The skin in the previous type is fine, and it is possible to pinch up with the fingers a little portion of it; but in these individuals none but a large fold of skin can be picked up, as it is so coarse. Speaking generally, persons of this class appear flabby and heavy-looking; they are apathetic, have little muscular power, and are soon tired. The vascularity of their tissues appears to be impaired, and leads to certain peculiarities of parts" (p. 87).

We now have to define and illustrate what is meant by the term "a similar series of kinetic action." This is a series of movements alike in time, and in quantity, to a second series. The use of the term involves the consideration of two series, and takes cognizance of the attributes time and quantity.

The term "series of movements" may be qualified in other ways than by the adjective "similar;" we may have a *good* series of movements. In an idiot the series of movements following upon the sight of an apple are not good; they are not such as follow in a healthy child from the sight of an apple.

If we measure the size of the head and different parts of the body of the idiot, we shall probably find that the proportions of such measurements are not good; they are not according to the proportions established by anthopometry as the normal for the age of the child. Mr. Roberts* has constructed tables indicating such proportions for normal children. If, further, we compare the series of movements observed in the idiot, with the normal series of movements, we shall probably find the kinetic series as defective as the series of measurements. The child's movements may be recorded by the experimental method,† and analyzed according to the principles of analysis,‡ and thus compared with a normal series as ascertained by previous experiments.

Many other considerations as to the attributes of a function might be presented as modes of expression, but sufficient has been said to illustrate the principles of analysis advocated in this work.

Some care has been taken to describe the attributes of trophic and kinetic functions, in the hope that,

* *Op. cit.*   † See chap. xix.   ‡ Page 256.

in studying the physical forces affecting motor action, a method may be found by which we may gain further knowledge of the processes of nutrition. Working towards this object, we proceed to show how the expression of development and other conditions may be described by series of movements and trophic actions; this will principally be demonstrated by examples.

Examine kinetic and trophic combinations in an infant. Is there not as much expression in the kinetic activities and successions as in the trophic actions, variations, and growth?

We say a body is well proportioned, or ill proportioned, large headed, small bodied, big mouthed, etc. All are expressions indicating the results of proportional development; the development or growth is the result of a series of trophic actions; the series may be well proportioned according to the normal. So the series of movements may be the normal or not. If the trophic series be abnormal the kinetic series is apt to be abnormal. In an infant, ill-proportioned in its head and body, the movements, combinations, and series of movements, and the reflex actions, are apt to be abnormal. The trophic abnormality is an expression that probably the movements and reflexes will be abnormal. In an idiot the movements are automatic in uniform series, in place of being varied as in health; the trophic action also occurs in abnormal series producing abnormal growth.

Expression may be described by series of movements and trophic actions. A series or succession of movements may be highly expressive. When we see a child turn his head towards an apple, extend his shoulder and elbow, then flex wrist and fingers till they grasp the apple, we observe a succession of movements expressing to us that the child sees the apple, and desires to have it; the expression is a series or succession of movements. When we compare the body of an infant one year old with a child aged three, the difference of their bodies in size, weight, proportional development, etc., indicates or expresses the difference of age. Some modes of expression are by kinetic function in the subject, others by its trophic functions. The expressive movements described in the child taking the apple are its kinetic functions; the points of difference between the body of the infant and the older child are its trophic manifestations.

Now, most of our biological works deal mainly with the description of the material presence of the subject, *i.e.* its trophic functions. Say the subject of the biologist's description is the growth of the common bean. The seed swells, each cotyledon enlarges, the radicle of the embryo elongates by growth, and protrudes from the opening in the seed covering called the micropyle. Then the covering is burst, the plumule protrudes, its hypocotyls bend towards the light, the leaflets of the plumule open out, etc. If carefully analyzed, the process of development described briefly will be found to consist of the enumeration of a series of trophic and kinetic actions of the parts of the subject. "The seed swells;" this is due to "enlargement of each cotyledon" (trophic). "The radicle grows and protrudes" (trophic action). In this example of growth, the ratio of the quantity of growth in the testa and the radicle is not constant as germination proceeds. Growth occurs more rapidly in the radicle than in other parts. The covering, or testa, bursts as a secondary result of growth of the cotyledons; the rupture is due to the absence of equal proportional development in embryo and testa. The plumule protrudes; rapid growth now occurs here (quantity of trophic action).

We may now inquire more closely as to each of these trophic processes. It is obviously desirable whenever we speak of a function, to know exactly what is the subject displaying that function. When we speak of the trophic function displayed in a bean as the radicle elongates, we do not mean to say that all parts are nourished to cause the radicle to grow, though this may be true: clearly we mean that the radicle displays the trophic function; the radicle is the subject, the vital action in it is said to be "trophic action." The question now arises, is this subject one and indivisible, or is it an assemblage of cells? If the latter, do all the cells composing the radicle go through a similar series of trophic action?

The regular circumnutation of the radicle involves numerous minor subjects, the cells of the radicle undergoing a regular series of trophic actions. It is also true that the nutrition produces partly trophic, partly kinetic, expression (see p. 283).

We will compare descriptions of growth and descriptions of movement. It is very desirable to know how to give scientific descriptions. A description may be given of the body of a tadpole and the body of a frog; the difference between the two is the growth or development of the individual. The description would be much more perfect if all the stages of the growth were given; the greater the number of "stages of growth" given, the fuller and more complete the description. Now contrast this with the description of the movements of the hand. We may observe the open "straight extended hand," followed immediately by the "convulsive hand;" we thus note two passive conditions, or postures, states of quiescence: the difference between them is an indication of the movement that has occurred. The description might be given by stating all the movements that actually occurred, flexion of all the joints of all the digits with adduction of the thumb upon the palm.

Again, in each of the examples the description might be given with great exactness and fullness, by recording the combinations and sequences of trophic and kinetic action as they occur, in time, quantity, and kind.

The object specially in view, in thus insisting on such methods of description, is the attempt to bring biological processes within the range of calculation. And it is also desirable to have a common form of expression of trophic and kinetic energy, so that we may more easily apply the same principles of analysis to both functions. It may be—I think it is true—that external forces, or the environment, do often bring about the special combinations and sequences of trophic action and kinetic action; and I imagine that here we have an important factor in the process of evolution of the individual and of the species.

Forces afferent to the subject may alter the succession of a kinetic series or the succession of a trophic series.

A child is at play, showing us much spontaneous movement and many reflex actions; speak to the child—the sound of your voice totally alters the character of his movements; the series of his kinetic actions is altered by the afferent force, your voice.

If a branch of vine be placed in darkness it will grow in abnormal porportions, with long internodes and small leaves, but will bear fruit. The proportional trophic series in the vine is altered by the action of light or darkness. It appears to me that often it is easier to observe this in the kinetic functions, than in the trophic functions. I believe, also, that very often in organic nature, plants and brains, a trophic action—indicated by permanent impressionability—does occur upon each display of kinetic function, and that this is the explanation of movements becoming automatic.

If this theory be true, it may be put to the test by my experimental methods. I believe that a subject may, as the result of its vital properties and nutrition, display the two functions $T =$ trophic action, $K =$ kinetic action; but, however much these two may vary, $T + K =$ the total function. Probably in most cases the time and duration of both functions are alike; hence we may use either T or K to indicate time and duration of the total function of the subject. The element or attribute "time" is thus made an essential part of the description.

In an investigation as to the development or evolution of a child's brain, we may proceed as follows, investigating only its development in growth, the changes in its material structure; at different ages its size, weight, anatomy, histology, chemistry, etc.

Again, we may investigate its kinetic functions by recording nerve-muscular signs. We find, at first, purely spontaneous movements, automatic, and not readily inhibited or controlled by light and sound. At three months old these movements are co-ordinated upon stimulus by light, later by sounds also. At this still early age, the special combinations and series of movements following upon a stimulus are uncertain, not uniform upon similar stimulations. Thus food placed in front of a new-born infant produces no uniform series of movements of its hands; food placed in front of a boy of six years stimulates him by the sight of it to move his hands and eat the bread and butter, his movements being so uniform—movements of hands, head, mouth, jaw, etc.—that we may call them uniform or automatic because they occur so regularly, so uniformly. Has this automatic action resulted from external agencies?

Given a description of the brain of a child two years old, such as may indicate whether it be healthy, we may conduct observations thus. As regards its material structure we can gain but little direct evidence. We can measure and take note of the form and size of the skull, or brain-case, and any changes occurring during the period of observation; we can test the special senses, examine the optic nerve, etc. But these methods of examination give us but little idea of the real condition compared with the knowledge gained by observing conditions of sleep (trophic) and conditions of movement. The rhythmical successions of sleep and of the recurrence of appetite for food, are successions of trophic conditions, highly expressive of health; the movements of the child, the coincidences of movements, and their series upon stimulation teach us much more. It is, then, the coincidences and successions of trophic and kinetic action that give the best kind of expression.

In a coincidence of movements many parts or few may concur. In the growth of a body many or few parts may grow at the same time.

Now I will give some reasons why it is desirable to give descriptions in this form. We have ample proof that the coincidences of movements are influenced by external forces: the sight of an object and sounds can cause special series of movements, this action makes it probable that external forces regulate the series of trophic actions.

In studying examples of proportional development, as illustrated in the case of the twins Arthur and John, it will be said, no doubt, that the cause of their equal proportional development is heredity; and this may be true. The question may, then, be raised, What is heredity? If proportional growth, and a definite series of movements be due to heredity, we have in such series a direct expression of heredity in a measurable form.

A child is growing like one or other parent in features (trophic combinations) and in manner of walking, speaking, or writing, etc. It is the series of trophic and kinetic actions that express the heredity. This seems easier to admit with regard to trophic than with regard to kinetic series. It seems more in accord with common observation that a certain mode of growth should follow on given external conditions than that a certain series of movements should succeed.

In the growth of children, on the average, the proportion of growth of limbs and parts of the body, is similar in different subjects at the same age. Is this owing to the impress made upon the embryo (or its germ) and upon its ancestors? It is certain that afferent forces may alter this, increase size of head, or hands, etc.

In the plant the variations of proportional growth are easily affected by light (internodes) and by heat.

In the case of the twin brothers, the reader will, from general knowledge, probably be inclined to accept the view that Arthur and John are of equal proportional development owing to heredity, owing to the fact of their being twins, owing to all the events of intra-uterine life having been similar to each. Is it not probable that the same cause is connected with the similarity of the series of trophic actions, and the series of kinetic actions? It is possible to describe walking and speaking in terms of a series of movements.

This suggests that "If two or more living subjects be co-nourished during the period of development they will tend to 'similar proportional development' and 'similar series of kinetic actions.'"

Let us try and see exactly what it is that is thus compared.

The series of trophic actions in each subject corresponds, or is similar in the same periods of growth; only observation can prove this. Observation shows in Arthur and John that growth is similar in kind, and in quantity, in each successive period of time; it is the attributes time, quantity, and kind of growth that are similar.

Summary.—Any property or function may be described as having the three attributes—time, quantity, kind. This may be illustrated with regard to growth, or trophic action; and also with regard to movement, or kinetic action.

The attribute time includes the moment of action, frequency, duration of the active appearance, and of the quiescence or disappearance of the function. Time is the only attribute of totally dissimilar functions that can be directly compared. As to quantity, this is measurable, but some unit of quantity must be adopted; before a comparison can be made between two quantities, it is necessary to determine a unit of measure common to the two.

The "kind" of a function may, in some cases, be stated in terms of time and quantity.

The properties of inorganic things vary less than those of living subjects.

The consideration of the attributes of properties in two or more subjects raises some important problems. The time of growth, or movement, in three subjects, A, B, C, may occur separately in each or may coincide in any of the combinations A, B, C, AB, AC, BC, ABC: thus considerations of time lead to coincidences, combinations, sequences, series.

As to quantities of growth, or movement, in two or more subjects, we are now able to make ratios or proportions. *Proportional growth*, inasmuch as it concerns a proportion, must imply something about the quantity of the growth; it is only as to quantities that we can make a proportion, or ratio. The term *proportional growth* is applicable to two subjects: it concerns the ratio of growth; and denotes the proportion of the quantity of the growth in the one subject, to the quantity of growth in the other subject.

The term *equal proportional growth* implies that two ratios are equal. Two subjects, each presenting two similar parts for comparison, are referred to; the ratio of the quantity of growth in the two parts of the one subject, is said to be equal to the ratio of the quantity of growth in the two corresponding parts of the other subject.

When growth is found to be good or bad in one part of a subject, it often happens to be similarly good or bad in other parts; this is termed the principle of *similar development*. The similarity is as to kind. In some cases this similarity as to kind is explicable, in other cases it is empirical. This principle is the basis of what are termed the "temperaments, or diathesis," in medicine.

One object of the analysis given, and the enunciation of principles in this chapter, is to show the usefulness of analogy between a series of movements and a series of trophic actions. A series of movements may be compared with the normal, as easily as a series of measurements is compared with fixed tables of the normal measurements. The expressions of development may be described in terms of series of kinetic and trophic actions; this is illustrated by examples.

Probably it may be demonstrated that forces afferent to the subject may bring about special combinations and sequences of trophic and kinetic action. I imagine that here we have an important factor in the process of evolution of the individual and of the species. A subject, as the result of vital action, may display two or more functions—say, trophic and kinetic action; the relative proportion of the two may vary in quantity—thus the total function of the subject may appear to vary in kind: all this may result from the action of afferent forces.

Some of the principles concerned in heredity may be illustrated by the principles enunciated in this chapter.

CHAPTER XVII.

ART CRITICISM.

Art teaches the physiologist; all men can study expression—Bulwer's opinion—All expression of feeling is by nerve-muscular action—Importance of such studies—Expression of mental states—Hand and face specially indicative of the mind—Venus de' Medici, the nervous hand—Diana, the energetic hand—Composition—Etruscan drawing—Cain at Pitti Gallery, the hand in fright—The Dying Gladiator—Writings of Camper; his descriptions of expression—Antony Raphael Mengs—Study of nervo-muscular action—Weakness should not be expressed in place of beauty—The free hand—The object of figure composition in art—Fixed and mobile expression—Principles of analysis.

As I gladly acknowledge my indebtedness to art for many suggestions as to modes of expression and the principles involved therein, I may perhaps be excused from the charge of rashness if I, though not an artist, venture to suggest some principles for the guidance of art representations of the emotions and other general conditions in man. The principles by the aid of which it is proposed to criticize art works, are those described in chaps. v. and ix. Ancient art has recorded indelibly in sculpture and on pottery the modes of expression seen in former times; the writings of Camper, Lavater, and Le Brun have taught much as to expression: and we shall inquire presently how far modern and contemporaneous art complies with the high requirements of the principles of expression in illustrating conditions of the body and the mind.

It is not necessary to speak here of artistic technique and execution; even without special knowledge on such matters any observer who thinks for himself and can compare the expression of a statue, or a painted figure, with living men and women, may form his opinions. If the principles for the analysis of expression contained in this volume be true and capable of wide application, they should be applicable to the criticism of artistic representations of expression. In framing those "principles" the analysis and comparative study of paintings and sculpture gave much help, as well as the analysis of examples in life; it is, then, not unnatural to apply the principles thus framed and enunciated to art criticism. All kinds of postures are produced by the action of the central nerve-mechanism, and, being the direct outcome of its function, are indices of its condition, and, as such, are worthy of study by observation, description, and analysis. Many admirable treatises have been written on expression, describing, in such terms as are above referred to, the motor outcome of those brain conditions whose mental manifestations are the emotions.

John Bulwer gave his descriptions in terms of

muscular action ("Criticism on Art," p. 5): "More strange yet, that no Artists should have made this the subject of their orations, but should have all to this Day, either turned their discourse to the structure only of the Humane Fabrique, the perfections or Symmetry of the Body, or the excellency and antiquity of the Anatomique Art, or the Encomiums of the Antient and moderne Anatomists; whereas nothing could have set a greater glosse upon the Art, or have bin more glorious and honourable, than together with their Dissections, to have inriched their discourse with a relation of the Essence, Regiment, and properties of the Soule, whose well-strung instrument the Body was; . . . for, what is more easie than to discerne the parts manifest to Sense, and the fidelity of a Ocular assurance? that are so subject to our touch, that in the semblances of those motions wrought in the parts by the endeavour of the Muscles, we may not only see, but as it were feele and touch the very inward motions of the Mind; if you aske what delight will hence acrew to the understanding? What is so delightfull as to know by what kind of movings those varying motions and expressions of the Head and Face are performed? . . . Wherefore we will think it a thing worthy to be corrected with the whip of Ignorance, if any rashly plunge himself into the Muscular Sea of corporal Anatomy, or of the outward man, without any mention of the Internal man, since the Soule only is the Opifex of all the moving of the Muscles, whose invisible Acts are made manifest by their operations in those parts into which they are inserted. . . . If they are contented to allow me to have bin the first that by Art endeavoured to linke the Muscles and the Affections together in a new Pathmyogamia . . . I ask no more."

"All the outward expressions we have or can make are performed by *motion*, and therefore signify the *affections* of the mind, which are *motions*; the *moving* of the instruments and parts, answering in a kind of semblance and representative proportion, to the *motions* of the mind" (p. 4).

"The Eyes of Man are the most cleere Interpretors of the affections of the mind; wherefore since they were for that purpose to be endued with a voluntary motion, and all motions are performed by Muscles, therefore the great Architect gave muscles to the Eyes, whereby they are most swiftly moved according to the inward motions of the mind, whence the Eye by the Philosopher is said to be the most moveable part of our Body, by which advantage it hath more opportunities to express the motions of our mind" (p. 166).

All the descriptions of expression given by Bulwer are muscular movements or nerve-muscular signs.

All expression of feeling is effected by muscular action, whether it be by words, by facial movement or gesture, movements effected by voluntary muscles; or expression may be produced by dilatation of the pupil, erection of the hair, or disturbed action of the heart, these being due to the conditions of involuntary muscular fibre.

Examples may easily be given showing how we commonly judge of the state of a man by muscular conditions. Note the stooping attitude and spiritless gait of a tired man as compared with that of the same individual when rested and refreshed. Incipient intoxication is indicated by a reeling gait, unsteady hand, and muscular tremor. Expression may be indicated by the position of the head: it is seen firmly upright in defiance, drooping in shame; it is commonly held on one side in nervous women, and girls convalescent from chorea, a good example of an asymmetrical gesture. The artist's brush or pencil, the sculptor's modelling tool and chisel, the pianist's and violinist's finger-touch, indicate the training and actual condition of the working of his brain. The educated and refined singer trains and refines his whole mind, *i.e.* his brain, and is well aware that his "whole soul," as he may express it, comes out in the action of the muscles concerned in producing his song and musical notes.

In the infant the condition of the nerve-system is best recorded in terms of nerve-muscular phenomena. It laughs and is playful; reflex action is well marked when a finger is placed in the child's hand or mouth; the eyes are moved and directed towards any object looked at; these are conditions of healthy action. It is well known that in the convulsive state the fists are often closed with the thumbs turned in.

All these examples of expression are nerve-muscular conditions; the movement, the attitude, the gait, result from states of the brain or spinal cord.

In the observations to be referred to, examples are chiefly drawn from the ocular and facial muscles, and those of the upper extremity.

If anything can aid our studies of man, the matter becomes of interest to several classes of writers, to all those who study the body of man as indicating the activities of his brain or mind, and as giving the knowledge of the means whereby the idea of his mental states and feelings may be expressed. If the matters discussed in this work are of use in this direction, they concern not only the physician but also the artist.

It is the work of the painter and sculptor, to express by form and posture, the conceptions he may wish to produce of the condition of men and women, in certain conditions of mind, states of strength and weakness; expressions of mental and physical pain, states of rest or repose, feminine coyness and defiance: the poet has to describe all these things in words. Clearly, also, it is a mistake for conditions of the limbs characteristic of disease to be used as mere expressions of feeling, unless the feeling be the result of disease.

There is, then, in the subject before us, conditions of the muscles expressive of the states of the nerve system, a field for observation and description, in which the artist and the physician may work together, observing and analyzing with as much exactness as may be, the modes by which the varying conditions of the brain and mind are indicated to our eye, and may therefore be described by words, or by drawing, or sculpture. We must study man in all aspects of the case, and when we see in the face, limbs, or body indications of his brain or mental condition, we should analyze and describe—first, the position of features and parts as we see them, then the muscles which produce these positions or movements, knowing that the muscular condition which has produced the movements or positions is the result of the state of the corresponding nerve-centres of the brain. It has been said that a man's face is the index of his mind, and this is true, for all the varying changes of expression in the face (except those of colour) are due to changes in the facial muscles, and these solely depend upon changes in nerve-cells.

The muscles of the hand and face are probably the most specialized as the agents, and indices of the mind. These muscles suffer most commonly from injury or disease of the brain. Another reason for speaking of the hand as specially indicative of the brain condition is because the hand has a large number of small muscles, capable of performing delicate actions, and bearing slight weights. The muscles of the hand are particularly under the guidance of the brain.

In 1879, when visiting Florence, it struck me that the posture of the hands of the Venus de' Medici was similar to the posture often seen in a nervous child when the hands are held out.

The hand posture of the Venus is similar on either side to that described as the "nervous hand" (see chap ix. p. 163).

296   PHYSICAL EXPRESSION.

It is a posture often represented in drawings and in sculpture. The wrist is slightly flexed or bent, the knuckle-joints are moderately extended

Fig. 32 — Venus de' Medici.

back beyond the straight line of the metacarpus; the fingers are slightly flexed; the thumb is extended backwards, and often slightly separated from the fingers. This spontaneous posture I have seen in hundreds of cases, usually in females of nervous, excitable temperament; in nervous children, bad sleepers, etc. In a child in whom this posture is usually seen when the hand is held out during the day, it is not seen when at rest. If, in such a child when asleep, the hand be held out by the wristband during sleep, the hand naturally falls into the posture of "the hand in rest."

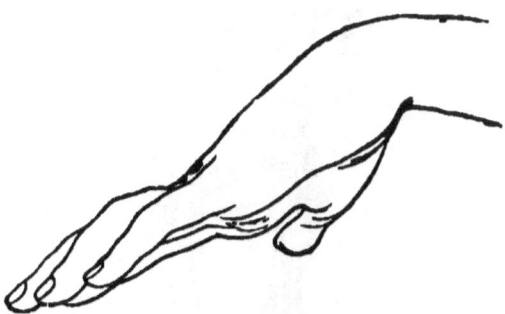

Fig. 33.—The Nervous Hand.

The most essential character of this nervous hand, is the extension backwards of the knuckle-joints and the thumb; this may be seen in some of the knuckle-joints only, as is well illustrated in Fig. 35, where each hand of the seated figures, if not engaged in holding a shield or bowl, presents one or more knuckles bent back beyond the straight line of the hand.

At the British Museum we have the statue of Diana next to that of Venus; the figure of a strong energetic woman in contrast with the figure expressing nervousness and coyness. In the Diana, the head is erect, the advanced right foot gives an expression of firmness. Now examine the hand

Fig. 34.—Diana. British Museum.

postures: the right is grasping a spear; the left arm hangs by the side, under the influence of gravity, but the hand is free, it is not engaged in doing anything, but its posture is the representation of the outcome of brain action only. As to the posture of this free hand under the influence of that condition of brain which makes a woman energetic, it is that described in chap. ix. as "the energetic hand;" it is often seen in children eager to answer a lesson in class and holding out their hands, or in a child running to a friend; it is commonly seen in an orator or preacher. The wrist is extended backwards, the fingers and thumb are all in moderate flexion. I believe this posture of the hand is common as the outcome of an active, energetic condition of the mind, and that the brain condition, which causes "an energetic condition of the mind," causes also at the same time, in many cases, the energetic hand. If the artist had so composed his figure that the left hand, as well as the right, had been represented engaged in holding some object, we should have had less expression of the woman's mental energy. The artist's skill appears in the composition in thus indicating by the right hand the physical strength, in the left hand the state of mental energy.

Fig. 35 is the photographic copy of an engraving by Mr. Kirk, in his work, "Outlines from Figures upon Greek Vases, etc., of the late Sir William Hamilton, MDCCCXIV."

"Plate I. represents a festival in honour of Bacchus, and consists of both sexes, who seldom or never were together except in these feasts." As already noticed, all the hands of the seated figures, where they are not engaged in holding some object,

300   PHYSICAL EXPRESSION.

present over-extension of the knuckle-joints, and

Fig. 35.—Feast of the Gods.

this is an essential element of the "nervous

hand." Now observe the hands of the Genius, who is not a partaker of the feast, and is not affected to nervousness. His hands are in the posture of energy; he alone of all the figures has the wrists extended. This contrast of the hand postures in the plate leads us to note that, just as nervousness and energy are the antithesis of one another, so "the nervous hand" and "the energetical hand" are antithetical postures (see chap. ix.).

Now let us look at the Cain in the Pitti Gallery, Florence. The whole figure expresses horror or mental fear. Each hand is free or disengaged, and in similar posture. The wrist is extended backwards as in the energetic hand of the Diana, indicating the large amount of nerve-force going to the muscles of the arms. This posture could serve no useful purpose to the man— it is not an act performed for the sake of doing something; it seems to be only the result of the force coming off from that condition of brain of which the "mental manifestation" is horror or fear. Here, as in the Venus, both hands are in similar posture. Analogous hand postures are seen in the members of the Niobe group. In the Dying Gladiator we learn a different lesson. Neither hand is here free. All the postures of the composition of this figure are the representation of a man in mortal agony, whose urgent dyspnœa determines the position of the body and of the limbs, which are thus not left free or disengaged to be acted upon solely by the spontaneous action

302   PHYSICAL EXPRESSION.

of the brain. Sir Charles Bell,* in his critical analysis of this posture, drew attention to this point; he says—

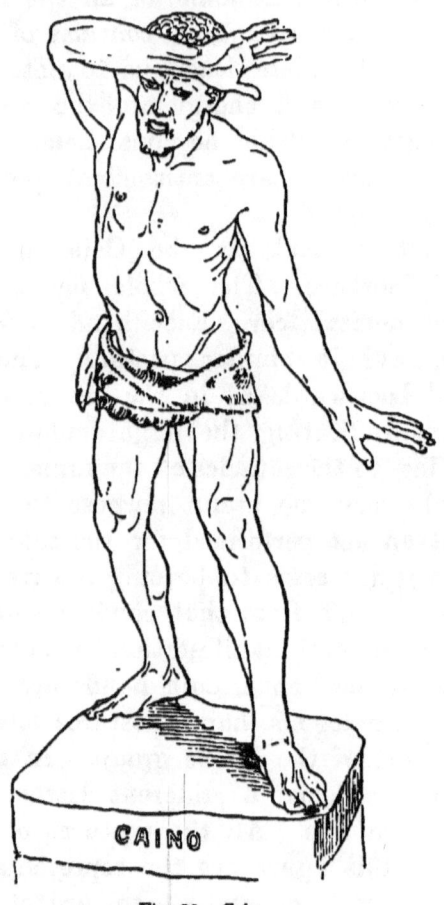

Fig. 36.—Cain.

"The Dying Gladiator is one of those masterpieces of antiquity which exhibits a knowledge of

* "Philosophy and Anatomy of Expression," 3rd edition, p. 195.

anatomy and of man's nature. He is not resting; he is not falling; but in the position of one wounded in the chest, and seeking relief in that anxious and oppressed breathing which attends a mortal wound with loss of blood. He seeks support to his arms, not to rest them or sustain the body, but to fix them, that their action may be transferred to the chest, and thus assist the labouring respiration. The nature of his sufferings leads to this attitude. In a man expiring from loss of blood, as the vital stream flows, the heart and lungs have the same painful feeling of want, which is produced by obstruction to the breathing. As the blood is draining from him, he pants and looks wild, and the chest heaves convulsively. And so the ancient artist has placed this statue in the posture of one who suffers the extremity of difficult respiration. The fixed condition of the shoulders, as he sustains his sinking body, shows that the powerful muscles, common to the ribs and arms, have their action concentrated to the struggling chest. In the same way does a man afflicted with asthma rest his hands or his elbows upon a table, stooping forwards, that the shoulders may become fixed points; the muscles of the arm and shoulder then act as muscles of respiration, and aid in the motion of the chest during the heaving and anxiety which belong to the disease.

"When a man is mortally wounded, and still more if he be bleeding to death, as the gladiator, he presents the appearance of suffocation; for the want is felt in the breast, and relief is sought in the heaving chest. If he have at that moment the sympathy and aid of a friend, he will cling to him, half raising himself, and twisting his chest with the utmost exertion; and while every muscle of the trunk stands out abrupt and prominent, those of the neck and throat, nostrils and mouth, will partake the excitement. In this condition he will remain fixed, and then fall exhausted with the exertion; it is in the moment of the chest sinking that the voice of suffering may be heard. If he have fallen on the turf, it is not from pain, but from that indescribable agony of want and instinctive struggling, that the grass around the lifeless body is lodged and torn."

In the figure of " Hercules at Rest " the position of the limbs is mainly determined by gravity; the figure presents in its build the signs of gigantic strength, but there is little or no expression of mentation.

The significance of the action of muscles as indicating brain conditions has long been dwelt upon by writers. Camper, who wrote in 1821, has shown how the Laocoon presents evidence as to how deeply the ancients had investigated the influence of pain as expressed in the figure and the muscles. In this group, " not merely does the face, but the arms, legs—in short, all the muscles of the body indicate anguish." Further on he quotes from the words of Paulo Somazzo's work, " Dell' Arte della Pittura," published 1531, in which he describes the influence of the passions upon the muscles of the face, and still more minutely the different postures and contortions of the body. Camper there complains that authors have usually either confined themselves to appearances, or have

Fig. 37.—Hercules at Rest.

"reasoned metaphysically concerning the operations of the mind, without attending to the physical causes of the changes produced by these operations, but in my opinion [that is, Camper's] speculations concerning the manner of the soul's working, or concerning the seat of the soul, are of no use to the artist. These belong to metaphysicians, who, by the way, lose themselves in a labyrinth of terms, or words with no definite meaning, without having in the least explained the action of this immortal principle upon the corporeal and mortal frame." Camper proceeds to give examples of the conditions of the muscles as indicating conditions of the mind, and then says, "the observation deducible from these effects is, that in every emotion of the mind particular nerves are affected; consequently every painter ought to make himself acquainted with the construction and connection of the nerves productive of these changes."

Camper * (p. 134) gives the following illustrations of expression, which he describes in terms of nerve-muscular signs thus (the figures are borrowed from his work):—

"Contemplate first the *placid* countenance (Fig. 38). Every feature is at rest; no one muscle is brought into particular action; all are in a state of repose, without appearing relaxed or inert. There is a tranquillity in the eye void of languor, and the lips are in unconstrained contact.

"Let us suppose something to present itself which excites a degree of *surprise* or *wonder* (Fig. 39). The intercostal nerves are immediately affected, and act upon the third pair; hence the eyelid is opened, and the eye stands motionless in the socket. The same nerve acts upon the eighth pair at the same

* Camper, Works, translated by T. Cogan, M.D., 1821.

time; respiration is suspended, the free motion of the heart is impeded, and the mouth is opened, as the maxillary muscles destined to this purpose are affected; but as these act alone upon the lower maxilla, the teeth are not discovered. The hands are extended, and more particularly the fingers, from the action of this muscular plexus.

"The effects of *contempt* are very different (Fig. 40). The fifth pair of nerves are put in motion. Thus

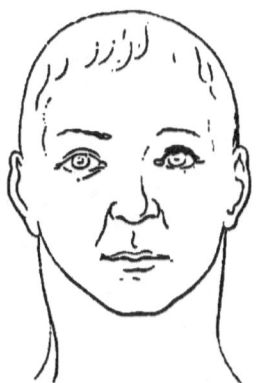
Fig. 38.—A countenance perfectly placid.

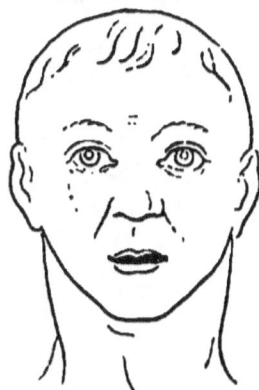
Fig. 39.—Expressing Surprise.

are the eyebrows drawn inwards and downwards; the mouth is firmly closed; but as the lower lip rises in the middle it becomes arched. The eyes are drawn sideways, the musculus abducens and adducens acting together by the force of habit. By making the head to turn towards the right, and the eyes toward the left hand, the passion is rendered more expressive.

(Fig. 41.) " In *complacency, friendly greetings,* and *tacit joy,* those parts alone act which have an immediate communication with the seventh pair of nerves. The angles of the mouth must never be drawn up alone, without other tokens of an incipient smile. Great care should be taken to avoid drawing the eyebrows inwards: an error frequently committed by the French in their portraits.

"*Laughter* (Fig. 42). In laughter all the effects produced by the former affection are greatly increased, and others are superadded. The whole countenance

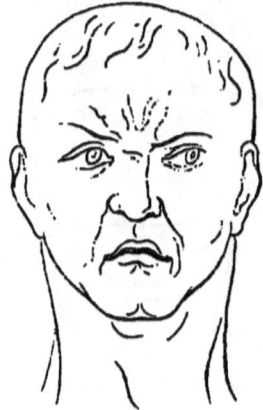

Fig. 40.—Contempt.

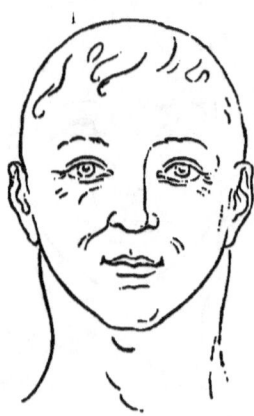

Fig. 41.—Complacency, Friendliness, Tacit Joy.

inclines forwards, but without the attention being fixed upon any determinate object. The outward edges of the orbicular muscles of the eye are contracted, producing wrinkles and folds around the eyes. The lips are opened by the action of the orbicular muscle, on the external sides; hence the teeth, particularly the upper, are made to appear; small wrinkles arise at the corners of the mouth, and the cheeks become fuller, etc.

"If you would add an arch, or a wanton look, place the eye sideways, and contract the upper eyelid expressive of a wink.

"In a *sorrowful* countenance (Fig. 43) the fifth pair of nerves are principally affected; the mouth is drawn downwards by the descent of the upper lip. To add *despair* to this emotion, the face must be made to look upwards and somewhat obliquely;

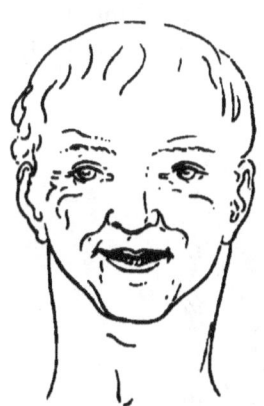 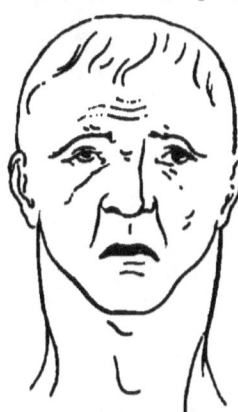

Fig. 42.—Laughter.   Fig. 43 —Sorrow.

the brow must be furrowed with wrinkles; and the middle of the eyebrows be drawn upwards.

"In weeping (Fig. 44), all the muscles which receive the fifth pair of nerves act in a very forcible manner. Hence the corners of the mouth are drawn downwards, the lower part of the nose upwards, the eyebrows descend, the eyes are nearly closed, and tears are pressed out of the lachrymal glands."

The following quotation is from the writings of Antony Raphael Mengs. He says:—

"By expression I mean the art of judiciously discovering the affections, by every sort of external signs. The union of the Soul with the body is of such a nature, that the emotion of the one, cannot happen without exciting a correspondent motion in the other. As the Painter ought therefore to represent his figures in action, he ought likewise to express in their appearance and in everything else, that situation and those emotions which the soul would produce in the body if it were really found

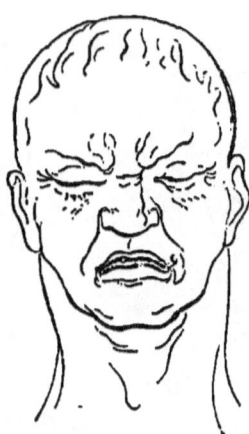

Fig. 44.—Weeping.

in that state; but since among these emotions, enter more or less, some which are forced, and others which are natural, some noble, and others ordinary, and of a thousand other manners; it depends therefore upon the Taste of the Painter, to know how to choose those which produce Beauty; and likewise to know how to produce it with due precision" (vol. i. p. 117).

"*Of the Proportions of the Human Body.*"[*]—Infinite are the descriptions of the proportions of the human body, but scarce any two accord. Those which I have read are not very clear, nor do I believe they can give to painters a just idea of the subject. Besides that, some authors have limited too much the combinations that could produce an uniform proportion in the figures. Others, and among whom is Albert Dürer, have explained a great number and variety of proportions; but they serve for nothing, except to those who would wish to imitate his taste. I shall therefore also say something on this subject, which might serve for all tastes, founding it upon nature and art.

"It is a general maxim to divide a figure in a determinate number of heads or faces; but this method will be good for sculptors only, and not for painters, who never see the heads just, because the perspective hides at least one third of the fourth superior part; and the width of the members cannot be measured with such exactness as they are measured by sculptors, because they would appear meagre and straight upon the plain surface, in opposition to what it appears by the perspective; because as we see all things with two eyes, we see the contour of things greater than the just diameter; and this happens in nature as well as in statues, but not in painting. The ancients also observed that, and therefore we see, that their bas-reliefs are thicker than their statues.

[*] Mengs' Works, vol. ii. p. 169.

"Painters have occasion to use variety infinitely more than sculptors, and of course have less subjections. Raphael, in a certain sense, only multiplied the taste of the ancients of the second order, by uniting it with a certain truth of which sculpture has not availed, either from rule or from taste, of all sorts of proportions, without being able to decide if one were better than another; and I know some of his figures, which have little more than six heads and an half; a proportion which would not be sufferable in any other than Raphael.

"The structure of the human body has such a symmetry, that it gives the idea of its motion, and this concordance of members is such, that to be able to produce that effect, one has to observe, what is called correctness of design. I shall therefore proceed to treat of this succinctly, proposing that which one ought to do to obtain it.

" The figure which one would wish to make being determined, one may design the head of the size one likes, observing, notwithstanding, for a rule, that the smallest head a painting admits of, is the ninth part of the figure, and the largest is a sixth part; these two dimensions are the two extremes; the general size being of an eighth or a seventh part of the figure. The neck should then be made equal to half of the head" (vol. ii. ch. xii. p. 159).

Mengs held that the affections of man may be discovered by every sort of external sign; that an emotion cannot occur in man without exciting a corresponding motion. He points out that the descriptions given of the proportions of the human body do not agree.

Some points in favour of the study of nerve-muscular conditions of the face are illustrated by the passive, expressionless face which may be woke up, "lighted up," made to express "the whole soul in the face" simply by conditions of tension of the facial muscles resulting from the mental state. Thus often great and most pleasing beauty is seen in faces unattractive when at rest. Conversely some faces are beautiful in their passive condition, but lack expression and interest when in action from mental work; such women talk but little. Surely from these two considerations it is suggested that the passive form and colour of a face are qualities not as great, not as much mind-indicating, as the mobile expressions produced by muscular tension.

Possibly we may for a moment be allowed to step out of our groove and see how education, and thoughts and habits of thought, and feeling, implanted in the individual can and do produce an effect on the individual's higher nerve-centres; for certainly direct observation shows that these principles, these forces or modes of force, alter the facial expression.

It is presumed that in composing and executing a work of art the object is to give expression and beauty to the whole. The expression, to tell its tale to the looker-on at the picture, must be more or less true to nature. The beauty of the picture is not a quality of the picture merely; it is such a quality in the picture as shall excite in the average or particular observer the sensation, feeling, or emotion of beauty. Sir Charles Bell* says, "In proceeding to define beauty, all that writers on art have been able to affirm is, that it is the reverse of deformity. Albert Dürer so expressed himself. If we intend the representation of beauty, then let us mark deformity, and teach ourselves to avoid it. The more remote from deformity, the nearer the approach to beauty." Bell tells us, on the next page, that Leonardo da Vinci searched for ugliness.

I would submit the proposition that the posture of every member of the body, in a drawing or sculpture, should be such as would in nature be produced in the subject under the circumstances represented or supposed in the composition.

Again, it is not allowable to use a posture at haphazard. The nervous hand should not be used except when the subject is weak with some degree of excitement or irritability; it should not be used as a posture merely to represent beauty, and should seldom be represented in men. It should never appear in a subject coincidently with signs of strength and energy; a nervous hand at the end of a strong arm would produce a most curious and grotesque appearance.

A head posture with slight rotation and inclination is appropriate to a subject directing the head towards an object on the side of rotation below the level of the head, or to a female subject with slight

---

* *Op. cit.*, p. 21.

weakness, but is inappropriate to sternness and defiance.

A not uncommon hand posture is the closed fist, or some modification of the convulsive hand, but this is inappropriate to the expression of rest. A slight modification of the convulsive hand is expressive of slight "shock" or terror, as in the piece of sculpture called "You Dirty Boy;" but "the hand in fright" is, I think, the outcome of terror.

It is to the free or disengaged hand that we must look for examples illustrating the condition of the brain which governs it. If the muscles be employed in some definite work, such as holding an object, or in an act of manipulation such as sewing, then the movements are directed to accomplish the aim attempted, and are not simply indicative of the condition of the brain, as may be the case with the free hand when unconsciously expressing the mental condition by gesticulation. When, on the contrary, the hands are left free and disengaged, as the hands of the orator, which unconsciously express by their position, or movements, the general mental state of the speaker, we have in these muscular movements an expression of the man's mind. It is as reasonable to look for the state of the mind to be expressed in the position and action of the hand engaged in definite voluntary, purposive acts, as to look for expression in the face when the sun is shining full in the eyes, or the lips are engaged in eating, or moved with the other movements of dyspnœa. Still it is true that in either case the manner of performing the act may be indicative of the mental state, but the muscles of the face or hand are not there engaged in expressing the mental state.

In art at the present day we but seldom see the hands represented as disengaged; usually they are painted or sculptured holding some object, or resting on some part of the figure; such are hands engaged or resting from labour, or performing some act of toil, not engaged in expressing the action of the mind.*

As to the average figure-compositions produced in the present time. When expression is desired, is care always taken to give to the figure its due form, proportions, and signs of development; and then so to compose the postures, and results of movements or other indications of movements, as to express something of the brain condition and mind of the living subject?

A thoroughly educated and able artist so draws his figure as to show thereby what kind of man he represents, and by his nerve-muscular signs shows us the mood of the man at the time of representation on the canvas. It is the postures and results of movement that directly represent the state of mind at the time.

Do most of our annual figure-pictures represent brain action? In order to give the highest effect to the representation of mobile expression, some parts of the figure should be left free and disengaged, especially hands and face. How common it is to

* Examples of the disengaged hand are seen in the statues of Cain in the Pitti, Florence; the Venus de' Medici; and the Diana, British Museum.

see the hands engaged in holding some object, in supporting the body, in resting upon other parts of the body. Such hands are not free to express the emotions; they express no other conditions than fatigue, rest, sleep, etc., states of mental inactivity.

I think there can be no doubt that the mobile conditions of expression, being the direct outcome of brain action, are higher in artistic value than mere proportions and signs of body growth; mere representations by colour effects are, I suppose, the least intellectual of all art productions, but not necessarily, therefore, the least pleasing.

Principles of Analysis and Composition.—Certain principles, derived from scientific study, may be suggested to the artist as likely to be useful in the analysis and composition of figures; they are the same as those given in chap. ix.; but it seems convenient here to put them in the form of rules and propositions. We have, then, in analyzing or composing a figure, to consider the proportions of the body, the more permanent and fixed conditions of the body (coincident development), and also the mobile expression.

The "principles" are applied to the analysis of the typical postures in the tables in chap. ix.

I. "Anatomical analysis."—In any figure-drawing the posture of every joint and part should be considered, and it should be possible, not only for the artist to see each posture in his imagination, but the looker-on at the picture should see the postures in each portion of the figure intended to be expressive, so as to be impressed by their meaning and significance.

II. "Small parts contrasted with large parts."—The significance of a shoulder movement differs from that of a finger movement; and a movement of the eyes in the head differs in signification from a movement of the skull (see p. 189.)

III. "Consider the different relative postures seen in the large, and small joints."—In a quiet, even action of the different parts of the brain and mind, there would probably be a similar condition of all joints; moderate flexion of all parts is the condition seen in rest and in sleep. A departure from such condition is therefore noteworthy and expressive.

IV. "Collateral differentiation."—Consider the relative condition of posture in collateral parts, whether it be similar in all the fingers, etc. Probably a certain amount of difference in posture, or movements of the fingers, is highly expressive of brain conditions—it may indicate mind or absence of mind. Look at the hand postures of brain disease in Fig. 16, p. 130. It is the different postures of fingers that makes them look like "hands gone mad."

When a nervous hand has all the knuckles over-bent, the expression of nervousness is greater than when only some of them are over-extended with the wrist flexion, as seen in the seated figures in Fig. 35, p. 300.

V. "Symmetry."—This term, as applied in art literature and description, is usually employed to indicate that the form of the body is alike on the two sides of the body. In speaking of mobile expressions, the term symmetry is intended to signify similar movements, or similar postures as the results of movements. In the Venus the hands are symmetrical, though the arms are not symmetrical. Thus we speak of symmetry of action when the limbs on each side move alike.

VI. "Excitation of weak centres."—A certain amount of mental excitement is expressed by the over-action of small muscles. Thus the backward extension of the fingers in Venus, and the straight extension of the fingers in Cain, show mental excitement. In the face the action of the small corrugator muscles of the forehead shows mental excitement; contrast this with the animal spirits expressed by the action of the larger (zygomatic) muscles used in grinning.

VII. "General excitement or weakness."—If a similar condition of excitement or weakness affect the body generally the stronger flexor muscles prevail. Hence, in convulsion, the hand is clenched, "the convulsive hand;" the same action is seen in strong passion.

If all the muscles are similarly relaxed, as in sleep, the stronger prevail, hence flexion results. In rest the hand is partially flexed.

VIII. and IX. "Analogy" and "Antithesis" are principles that will, I think, often assist in composition.

In a piece of sculpture exhibited at the Royal Academy, a child's hands were represented with the fingers flexed and the thumbs turned in—a posture very analogous to what would be seen if the child were going off into a fit. This analogy was not pleasing. In the admirable piece of statuary executed for Messrs. Pears and Co., and exhibited by them at the Paris Exhibition, the "Dirty Boy" does not like being washed; he rebels against it, and would escape if he could. The free hand is very expressive: the knuckles are somewhat flexed though the fingers are straight, and the thumb is strongly pressed against the index finger. This posture is so analogous to the convulsive hand as to express well the semi-convulsive condition of the boy.

Antagonistic conditions of the mind are probably often indicated by antithetical postures.

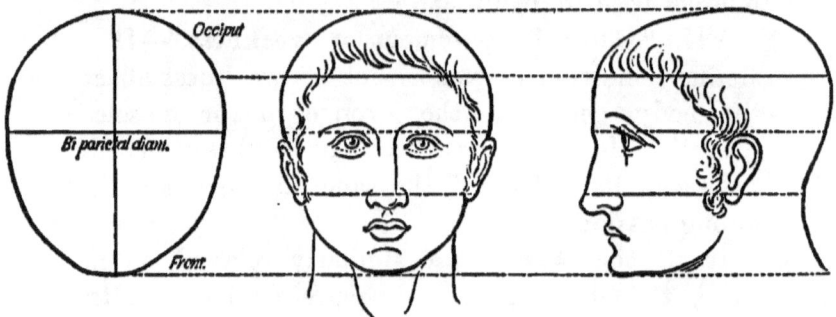

Fig. 45.—Showing Facial Zones.

CHAPTER XVIII.

LITERATURE.

Bulwer—Hartley—Gregory—Camper—Blanc—Marshall Hall—Tyndall—C. Darwin—Bibliography, with dates.

In looking over the writings of some authors on subjects cognate to that of the present work, there appears to be abundant evidence, in the few extracts here made, that some of them are in accord with the general principles expressed in this volume, although, of course, they are not enunciated in the same way, owing to the present advanced condition of knowledge in cognate matters. If it can be shown that the methods of inquiry now proposed are in unison with the thoughts of men long past, but whose works are still of value, it appears to me that the probability of useful results following from such work is strengthened.

Bulwer* says, "Being their motions, then, are diametrically opposite, so are their significations; for if we use, as we do, the flexion of the Head to show our assent, and that properly according to the universal intention of Nature; the contrary

* Bulwer, *op. cit.*, p. 56.

motion to that must as naturally imply *dislike*, or *dissent*, or *resentment.*"

"And who, I pray you, that is well versed in Philosophy, does affect to behold the cold effects of common Actions, without a Discourse of their causes and intrinsicall Agents, the Soule and the Muscles? Since that is familiar to Sense, and so by consequence to Beasts. But this is subjected to the Intellect, to-wit, the Internall Principal of man, wherefore we will think it a thing worthy to be corrected with the whip of Ignorance, if any rashly plunge himself into the Muscular Sea of Corporal Anatomy, or of the outward man, without any mention of the Internall man, since the Soule only is the Opifex of all the movings of the Muscles, whose invisible Acts are made manifest by their operations in those parts into which they are inserted. Not that any perfection or exact knowledge of this nature can be acquired; since the wisdome of the Creator in the fearefull and wonderfull structure of the Head is not yet fully found out, although it has beene sought after by illustrious men, with much Piety and Deligence" (Introductory chapter, p. 10).

"*Of the Muscles serving to the generall expressions, or most important motions of the Face or Countenance.*

"Many are the affections of the Mind that appear in the significant motions of the Face, even when the Bones are at rest; for whereas some parts of our skin are altogether immovable and pertinacious in their circumduction over the subjacent parts; other parts thereof versatile indeed, but they are not actuated with any voluntary motion: the skin of the whole Face participates of motion, which being voluntary, does necessarily imply the use of Muscles, by whose benefit those motions should be orderly and significantly performed" (p. 97, No. ii.).

"But in the *face*, it hath a publique and locall motion that is most Emphatically significant, wherein the perturbations of the Mind discover themselves, being moved in the Face by the streight annexion to Muscles, which are the organs of voluntary motion; for, Nature would have so ordered, that by the benefit of certaine Muscles working under the skin, and affecting the parts of the Face, (being all of them furnished in their originals with Nerves from the third Conjugation of Nerves that come from the Braine) Man with his very Countenance alone, should expressse all his Will, Mind, and Desire, when at any time it happened to be inconvenient or unlawful to open it in words at length" (p. 101).

"All the outward expressions we have or can make, are performed by *motion*, and therefore signifie the *affections* of the mind, which are *motions;* the moving of the instruments and parts, answering in a kind of semblance and representative proportion, to the *motions* of the mind" (p. 4).

"Now the Braine is the *Universall organ* of *voluntarie motion*, the great mysterie whereof is thus ordered. The *Braine commandeth* as soone as it hath *judged* whether the thing is to be avoided or prosecuted" (p. 13).

"You shall find in it that which I use to call the clock-work of the Head, or the Springs and inward

Contrivance of Instruments of all our outward motions, which give motion and regulate the Dyall of the Affections, which Nature hath placed in the face of Man; Being a New Light, and the first Irradiation which ever appeared through the Dissections of a Corporeal Phylosophy" (Dedicatory, p. 2).

"The *whole head* with its comprised parts, by the benefit of certain *Muscles* might be enabled to move, and by *motion*, to expresse the affections of the mind" (p. 51).

"Having resolved to trace the Discoursing Actions of the Head to their Spring and Principle upon which their outward significations depend; when I had passed the superficial parts, and digged a little more than skin-deepe into the Minerall of Cephalicall Motion; I came to the Muscles, the instruments of voluntary motion; or the instruments of those motions that are done by an earnest affection, that is, from an inward principle. The effects of whose moving significantly appeare in the parts moved, when by an arbitrary motion we freely reject or embrace things understood (not with our minde only, but with our mind and body both)" (p. 4).

"There hath not been any one who, *Data Operâ*, had undertaken a generall Survey and Cognomination of the muscles of the Body, as they are the necessary Instruments of all those motions of the Mind, which are apparently expressed and made manifest by the effect of their use and moving in all the parts of the Body; although more Emphatically, by those operations they have in the Head and the most remarkable parts thereof" (A. 3).

"When we lightly dislike, refuse, deny, or resent a thing, we use a cast up backward Nod of our Head, a motion diametrically opposite to the forward motion of assent; and this signification of the mind is performed by the extension of the Head" (p. 54).

Bulwer (pp. 56 and 54) suggests the principle of antithesis. He saw that movements, or nerve-muscular signs, are indices of the brain condition which accompanies mental states (see p. 4).

Hartley,[*] in his "Theory of the Human Mind," writes (p. 31): "Association not only converts automatic actions into voluntary, but voluntary ones into automatic. For these actions, of which the mind is scarce conscious, and which follow mechanically, as it were, some precedent diminutive sensation, idea, or motion, and without any effort of the mind, are rather to be ascribed to the body than the mind; *i.e.* are to be referred to the head of automatic motions. I shall call them automatic motions of the secondary kind, to distinguish them both from those which are originally automatic, and from the voluntary ones; and shall now give a few instances of this double transmutation of motions, viz. of automatic into voluntary, and of voluntary into automatic. The fingers of young children bend upon almost every impression which is made upon the palm of the hand, thus performing the action of grasping, in the original automatic manner.

[*] Edition by Joseph Priestley, F.R.S., 1775.

(*a*) After a sufficient repetition of the motions which concur in this action their ideas are associated strongly with other ideas, the most common of which, I suppose, are those excited by the sight of a favourite plaything which the child uses to grasp and hold in his hand. He ought, therefore, according to the doctrine of association, to perform and repeat the action of grasping upon having such a plaything presented to his sight. But it is a known fact that children do this. By pursuing the same method of reasoning, we may see how, after a sufficient repetition of the proper associations, the sound of the words *grasp, take, hold,* etc., the sight of the nurse's hand in a state of contraction, the idea of a hand, and particularly of the child's own hand, in that state, and innumerable other associated circumstances, *i.e.* sensations, ideas, and motions, will put the child upon grasping, till at last, that idea or state of mind, which we may call the will to grasp, is generated, and sufficiently associated with the action to produce it instantaneously. It is, therefore, perfectly voluntary in this case; and by the innumerable repetitions of it in this perfectly voluntary state, it comes at last to obtain a sufficient connection with so many diminutive sensations, ideas, and motions as to follow them in the same manner as originally automatic actions do the corresponding sensations, and consequently to be automatic secondarily. And, in the same manner, may all the actions performed with the hands be explained, all those that are very similar in life passing from the original automatic state through the several degrees of voluntariness till they become perfectly voluntary, and then repassing through the same degrees in an inverted order, till they become secondarily automatic on many occasions, though still perfectly voluntary on some, viz. whensoever an express act of the will is exerted."

"I will here add Sir Isaac Newton's words, concerning sensation and voluntary motion, as they occur at the end of his *Principia*, both because they first led me into this hypothesis, and because they flow from it as a corollary. He affirms, then, 'both that all sensation is performed, and also the limbs of animals moved in a voluntary manner, by the power and actions of a very subtle spirit; *i.e.* by the vibrations of this spirit, propagated through the solid capillaments of the nerves from the external organs of the senses to the brain, and from the brain into the muscles.' "*

Hartley refers to consciousness as an item in the voluntary character of certain movements. He also refers to the movements of young children as being at first automatic, then voluntary, giving out the idea that the study of infantile movements is a basis for the study of mind.

Gregory, in his work entitled "A Comparative View of the State and Faculties of Man with those of the Animal World," writes (p. 4): "Enquiries into the structure of the Human body have, indeed, been prosecuted with great diligence and accuracy. But this was a matter of no great difficulty. It

---

* Hartley, *op. cit.*, p. 39, Cor. 4.

required only labour and a steady hand. The subject was permanent; the Anatomist could fix it in any position, and make what experiments on it he pleased.

"The Human Mind, on the other hand, is an object extremely fleeting, not the same in any two individuals, and ever varying even in the same person. To trace it through its almost endless varieties, requires the most profound and extensive knowledge, and the most piercing and collected genius. But tho' it be a matter of great difficulty to investigate and ascertain the laws of the mental constitution, yet there is no reason to doubt, however fluctuating as it may seem, of its being governed by laws as fixt and invariable as those of the Material System. It has been the misfortune of most of those who have study'd the philosophy of the Human Mind, that they have been little acquainted with the structure of the Human Body, and with the laws of the Animal Œconomy; and yet the Mind and Body are so intimately connected, and have such a mutual influence on one another, that the constitution of either, examined apart, can never be thoroughly understood. For the same reason it has been an unspeakable loss to Physicians, that they have been so generally inattentive to the peculiar laws of the Mind, and to their influence on the Body.

"A late celebrated professor of Medicine in a neighbouring nation, who, perhaps, had rather a clear and methodical head, than an extensive genius or enlarged views of Nature, wrote a System of

Physic, wherein he seems to have considered Man entirely as a Machine, and makes a feeble and vain attempt to explain all the Phenomena of the Animal Œconomy, by mechanical and chymical principles alone. Stahl, his cotemporary and rival, who had a more enlarged genius, and penetrated more deeply into Nature, added the consideration of the sentient principle, and united the philosophy of the Human Mind with that of the Human Body: but the luxuriancy of his imagination often bewildered him, and the perplexity and obscurity of his style occasion his writings to be little read and less understood.

"Besides these, there is another cause which renders the knowledge of Human Nature very lame and imperfect, which we propose more particularly to enquire into.

"Man has been usually considered as a Being that had no analogy to the rest of the Animal Creation. The comparative anatomy of brute Animals hath indeed been cultivated with some attention; and hath been the source of the most useful discoveries in the anatomy of the Human Body. But the comparative Animal Œconomy of Mankind and other Animals, and comparative Views of their states and manner of life, have been little regarded. The pride of Man is alarmed, in this case, with too close a comparison, and the dignity of philosophy will not easily stoop to receive a lesson from the instinct of Brutes. But this conduct is very weak and foolish. Nature is a whole, made up of parts, which, though distinct, are yet intimately connected with one another. This connection is so close, that one species often runs into another so imperceptibly, that it is difficult to say where the one begins and the other ends. This is particularly the case with the lowest of one species, and the highest of that immediately below it. On this account no one link of the great chain can be perfectly understood, without the knowledge, at least, of the links that are nearest to it."

"The active principle is so vigorous and overflowing in a Child, that it loves to be in perpetual motion itself, and to have every object around it in motion. This exuberant activity is given it for the wisest purposes, as it has more to do and more to learn in the first three years of its life than it has in thirty years of any future period of it. But that lively and restless spirit, which in infancy seemed to animate everything around it, gradually contracts itself, as the Child advances in Life, nature requiring no more motion than is necessary for its preservation, and sinks at last into that calm and stillness which close the latter days of human life" (p. 61).

Gregory speaks of the human mind as an object extremely fleeting and ever varying. He believed it to be governed by laws as fixed and invariable as those of the material system. Gregory thought that an inquiry as to mind, should embrace considerations concerning other living beings in nature besides man. It would appear that one of the great difficulties that he saw in the way of the study of mind was the fleeting character of its signs, or manifestations. If those signs can be permanently recorded, and automatically enumerated, this objection seems greatly lessened.

Camper* thus wrote: "An oppressed, sorrowful, and melancholy person, lets his head sink downwards, or he supports it with his hand; the equipoise is no longer maintained by the muscles of the neck; that is, the nerves belonging to those muscles are rendered inert.

"A lively contented laugher, on the other hand, raises his head, and his breast is agitated. In the excess of his emotion, he places both his hands to his sides, as it were to support his body. At length his legs begin to refuse their office, and he would fall to the ground if the fit continued.

"A person in the impetus of rage, beats with hands and feet, stamps till the ground shakes under him; and his face is convulsed in a thousand forms.

"Deep reverence makes the tongue to falter, an inward trembling impedes the motion of the body; the most lively and expressive eyes are abashed, and look downwards; the heart flutters; if shame accompany this emotion, as is frequently the case, the face, neck, and breast are immediately painted of a crimson colour.

"It would be endless to particularize every emotion in a similar manner. The observation deducible from these effects is, that in every emotion of the mind particular nerves are affected; . . . orators and public actors have the superior advantage of giving the greatest force to the

* "On Painting, Sculpture, etc.," p. 129.

expressions of the features, by exciting requisite movements in the parts themselves."

Camper is one of the writers who fully recognized the significance of nerve-muscular signs as means of expression.

Sir Gilbert Blane * in his Croonian Lecture, delivered before the Royal Society, says (p. 233):

"So far as we know, either from actual observation, or from analogy, there does not exist in nature any such thing as absolute *rest;* for when we contemplate the motions of the earth and heavenly bodies, the various complications of the planetary revolutions in their rotation round their own axes, and in the paths of their orbits, in the irregularities arising from the disturbances of their mutual gravitation, and from the precession of the equinoxes, not to mention the influence of the innumerable sidereal systems upon each other, it may be affirmed, on incontestable principles, that no particle of matter ever was, or will be, for two instants of time, in the same place, and that no particle of it ever has returned, or will return, to any one point of absolute space which it has ever formerly occupied. Whether motion, therefore, can strictly be called an essential property of matter or not, it is, certainly, by the actual constitution of nature, originally and indefeasibly impressed upon it; and as rest does not exist in nature, but may be considered in a vulgar sense, as a fallacy of the senses, and in a philosophical sense, as an abstraction

* "Select Dissertations on several Subjects of Medical Science." 1822.

of the mind, it follows, that what is called the *vis inertiæ* of matter, is not a resistance to a change from rest to motion or from motion to rest, but a resistance to acceleration or retardation, or to change of direction. If it should be alleged, that any given particle or portion of matter is carried along by value of the motion of the planet to which it belongs, it may be answered, that the earth or any other planet is nothing more than a congeries of such particles, each of which must possess a share of the same energy which animates the whole mass."

" I have already acknowledged my ignorance of the manner in which *stimuli* in general operate, and that this must be admitted as an ultimate fact in nature. But the operation of the will through the nerves, seems involved in double obscurity; for as it depends on the nature of thought, it cannot be made a subject of experimental investigation. For this reason I shall decline the enquiry, as not being adapted to the ends of this Society: and it seems impossible for human sagacity to penetrate the connection of matter with sensation and volition. All such attempts have consisted of abortive and unsatisfactory inferences drawn from hypothetical assumptions. The properties of different bodies, in relation to each other, appear to be the only proper subjects of experimental reasoning; for, in their relation to the mind, they are only the effects, perhaps the remote effects, of their intimate nature upon the senses; and we may venture to affirm that human reason can no more fathom the connection of thought with the corresponding changes in the corporeal organs, than the eye can see itself" (p. 258).

"Not to mention the well-known effects of grief, fear and joy, which affect the whole circulation, there are certain passions and sentiments which produce partial and local effects.

"Fear produces debility, almost amounting to palsy. Courage and ardor of mind, on the contrary, adds to the natural strength. When the mind is agitated by some interesting object, and calls upon the body for some extraordinary exertion to effect its end, the muscles are thereby enabled, as it were by magic, to perform acts of strength, of which they would be entirely incapable in cold blood. In circumstances of danger, for instance, where life or honour are at stake, exertions are made in overcoming mechanical resistance, which seem incredible, and would be impossible, were not the mind in a sort of phrenzy! and it is truly admirable, in the economy of nature, that an idea in the mind should thus in a moment augment the powers of motion and inspire additional resources of strength, adequate to the occasional calls of life. The great increase of strength in maniacs, is also referable to the passions of the mind. These considerations would almost lead us to doubt whether or not the accounts we have of the great feats of strength ascribed to individuals in the heroic ages are fabulous or not" (p. 259).

"The other class of *stimuli* to be enumerated are the external. These consist in impressions made by outward bodies. They are either immediate, as in the case of those motions which are excited, whether by mechanical means, or by acrimony, directly and artificially applied to a muscular fibre; or they are remote, as in the various instances of sympathy, and in the case of those instincts which nature has implanted for the purpose of self-preservation in brutes, and in the early part of human life. I shall here confine myself to a few remarks on instinct, as the other branches of this subject have been fully and ably handled by those who have gone before me in this Lecture. There is a connection established between the impression of certain external bodies and the action of certain muscles, analogous to what has already been noticed with regard to the internal motions excited in vessels by the peculiar *stimulus* of their fluids, Nature having instituted certain habitudes between outward *stimuli* and the moving powers, whereby natural propensities are established equally necessary to the support of life as the internal functions. Thus, in a new-born animal, the first contact of the external air excites the act of respiration, and the contact of the nipple excites the act of sucking; both of which actions are absolutely necessary to the maintenance of life, and require the nice cooperation of a great number of muscles, prior to all experience. Actions of this kind are called instinctive, and differ from voluntary motions in this respect, that the latter are the result of memory and experience, whereas the former are the immediate effect of external impressions, in consequence of an established law of nature, and independent of consciousness. The actions of instinct and those of volition, nevertheless, run imperceptibly into each other, so that what was at first instinctive may afterwards come to be a matter of deliberate choice. The same muscles are the instruments of both, and they differ from the muscles obeying the internal *stimuli*, such as the heart, in this respect, that they are liable to fatigue, and thereby concur with the exercise of sensation and of thought, in rendering sleep necessary. There are no muscles, except those of respiration, of which the constant action is necessary to life, and which are void of consciousness in their ordinary exercise, but which are occasionally under the control of the will. The principal end answered by this power of the will over the muscles of respiration in man, is to form and regulate the voice.

"But though instinctive motions are in some cases convertible into those which are voluntary, we should be so far from confounding them, that the former are even compatible with the want of consciousness and sensation; for those animals which are destitute of brain and nerves, are capable of actions analogous to the instinctive. There is something very similar to this even in vegetables, as in the case of tendrils and creeping plants being stimulated by the contact of other bodies, to cling round them in a particular direction" (p. 260).

Blane (p. 258) taught that will and thought could not be subjects of scientific inquiry, because they could not be submitted to experimentation.

The properties of different bodies in relation to each other appeared to him to be the only proper subjects of experimental reasoning. If, then, we can show how, under the action of various forces, the parts of the brain concerned in thought act in relation to each other, we may obtain some knowledge as to what forces bring about that relationship, although we cannot affirm the connection of thought with the corresponding changes in the corporeal organs. We do not know much about heat, light, and electricity, except as to their display in corporeal subjects.

Blane then proceeds to describe certain means of expression of the emotions by motor signs. He considered that no sharp line could be drawn between instinctive and voluntary actions; he thought instinctive movements to be comparable with movements seen in the tendrils of plants.

Marshall Hall* said, "The particular circumstances embraced in an examination of the morbid states of the countenance are the changes induced in the *cuticular surface, the cutaneous circulation, the cellular substance, the muscular system, some particular features, and the general expression.*"

"§ 61. The state of *emaciation*, so important to observe and trace in chronic diseases, depends on the loss of cellular and muscular substance, and must be always distinguished from mere vascular shrinking.

"§ 62. The muscular system is principally affected by diseases attended with pain, languor, or paralysis."

* "The Principles of Diagnosis," vol. i. p. 31, par. 58.

"§ 64. Of the general expression of the countenance I shall rarely venture to speak. It affords an important and essential source of information in Dispensary practice, § 50, and assists the experienced physician in discerning the nature of the disease where the superficial observer sees only the general look of indisposition."

"§ 121. I employ the term attitude in a rather comprehensive sense, intending to embrace, under this head, the consideration of *the postures and motions of the body, the state of muscular debility, power, contraction, and motion, some particular actions, and the general manner of the patient.*"

"§ 125. Certain movements of the head, certain actions of the hand, and certain peculiarities of the general manner, also occur as characteristic of particular diseases, and will be noticed hereafter."

"§ 127. In healthy and undisturbed sleep, the usual posture is that on one side, the body being frequently inclined rather to the *prone* than to the *supine* position; the head and shoulders are generally somewhat raised, and together with the thorax, bent generally forwards; the thighs and legs are in a state of easy flexion. The position is apt to be changed from time to time, the person lying on one or other side alternately."

"§ 133. In the severe forms of *Typhus Fever* the position of the patient becomes more and more supine, and the actions more and more tremulous; from being able to retain the posture on the side, perhaps, the patient falls upon his back, with the lower extremities extended and sometimes with a tendency to sink towards the bottom of the bed ; the hands and arms are moved with effort and tremor, and at length there is constant subsultus tendinum. To this state, picking of the bed-clothes, or of flocci volitantes, delirium, or coma, is superadded."

"§ 135. As this position is occasioned by extreme debility, any change of posture is of favourable omen, as denoting a return of strength. The patient, perhaps, raises the knees, or puts the arms out of bed, or places them above his head. These movements are among the first symptoms of recovery. At length the patient is capable of supporting the position on the side—a certain mark of returning muscular strength, and an indication of a favourable change in the disease."

"§ 140. The form of tremor which I have described seems to depend on muscular debility, and perhaps on a morbid condition of the brain and nervous system. There is a kind of tremor of a more *spasmodic* character, which occurs from various causes, and which I shall notice towards the conclusion of the present chapter."

Marshall Hall studied the general appearance of his patients and their various modes of expression: the condition of nutrition, the attitude, postures and movements of the body, the movements of the head and hand. He described the position of the body in sleep, in typhus fever, and in conditions of debility, especially noting that in the latter condition tremor was often a symptom.

Tyndall[*] says in his lectures : " The matter of

[*] "Heat a Mode of Motion," p. 501, par. 722.

our bodies is that of inorganic nature. There is no substance in the animal tissues which is not primarily derived from the rocks, the water, and the air. Are the forces of organic matter, then, different in kind from those of inorganic? All the philosophy of the present day tends to negative the question, and to show that it is the directing and compounding, in the organic world, of forces belonging equally to the inorganic, that constitutes the mystery and the miracle of vitality.

"(723.) In discussing the material combinations which result in the formation of the body and the brain of man, it is impossible to avoid taking side glances at the phenomena of consciousness and thought. Science has asked daring questions, and will, no doubt, continue to ask such. Problems will assuredly present themselves to men of a future age, which, if enunciated now, would appear to most people as the direct offspring of insanity. Still, though the progress and development of science may seem to be unlimited, there is a region beyond her reach—a line with which she does not even tend to osculate. Given the masses and distances of the planets, we can infer the perturbations consequent on their mutual attractions. Given the nature of a disturbance in water, air, or ether, we can infer from the properties of the medium how its particles will be affected. In all this we deal with physical laws, and the mind runs freely along the line which connects the phenomena from beginning to end. But when we endeavour to pass, by a similar process, from the region of physics to that of thought, we meet a problem not only beyond our present powers, but transcending any conceivable expansion of the powers we now possess. We may think over the subject again and again, but it eludes all intellectual presentation. The origin of the material universe is equally inscrutable. Thus, having exhausted science, and reached its very rim, the real mystery of existence still looms around us. And thus it will ever loom—ever beyond the bourne of man's intellect—giving the poets of successive ages just occasion to declare that—

> " 'We are such stuff
> As dreams are made of, and our little life
> Is rounded by a sleep.' "

Professor Tyndall did not, in 1868, look hopefully forward to the possibility of an experimental inquiry as to the causes of mentation or the faculty of the brain to produce mind. I think the methods of inquiry proposed in chap. xix. are analogous, not to say copied from, the modes used by the physicists. It may be maintained that the work entitled "Heat a Mode of Motion" chiefly deals with the obvious expressions of heat in a manner very analogous to that in which we propose to study mind, by observing and experimenting with the forces producing "mentation." To illustrate this, quotations are taken from the "contents" of the volume referred to:—

I. Instruments; II. The nature of heat; III. Expansion; IV. The Trevelyan instrument; V. Application of the dynamical theory; VI. Convection of heated air; VII. Conduction; VIII. Cooling a loss of motion; IX. Law of diminution with the distance; X. Absorption of heat; XIII. Discovery of dark solar rays; XIV. Dew. These are matters concerning relations, time, quantity, visible phenomena, etc. Such are the proper subjects of scientific study, and such phenomena accompanying mentation are capable of experimentation. I think, then, that Professor Tyndall's remarks and his whole work sustain the advisability of the work proposed as a study of brain centres.

Charles Darwin gave a biographical sketch of an infant in *Mind*, 1877. The following extracts are illustrative of our subject :—

"During the first seven days various reflex actions, namely sneezing, hickuping, yawning, stretching, and of course sucking and screaming, were well performed by my infant. On the seventh day, I touched the naked sole of his foot with a bit of paper, and he jerked it away, curling at the same time his toes, like a much older child when tickled. The perfection of these reflex movements shows that the extreme imperfection of the voluntary ones is not due to the state of the muscles or of the co-ordinating centres, but to that of the seat of the will. At this time, though so early, it seemed clear to me that a warm soft hand applied to his face excited a wish to suck. This must be considered as a reflex or an instinctive action, for it is impossible to believe that experience and association with the touch of his mother's breast could so soon have come into play. During the first fortnight he often started on hearing any sudden sound, and blinked his eyes. The same fact was observed with some of my other infants within the first fortnight. Once, when he was sixty-six days old, I happened to sneeze, and he started violently, frowned, looked frightened, and cried rather badly; for an hour afterwards he was in a state which would be called nervous in an older person, for every slight noise made him start. A few days before this same date, he first started at an object suddenly seen; but for a long time afterwards sounds made him start and wink his eyes much more frequently than did sight; thus when 114 days old, I shook a pasteboard box with comfits in it near his face and he started, whilst the same box when empty or any other object shaken as near or much nearer to his face produced no effect. We may infer from these several facts that the winking of the eyes, which manifestly serves to protect them, had not been acquired through experience. Although so sensitive to sound in a general way, he was not able, even when 124 days old, easily to recognize whence a sound proceeded, so as to direct his eyes to the source. . . . The movements of his limbs and body were for a long time vague and purposeless, and usually performed in a jerking manner; but there was one exception to this rule, namely, that from a very early period, certainly long before he was forty days old, he could move his hands to his own mouth. When seventy-seven days old, he took the sucking-bottle (with which he was partly fed) in his right hand, whether he was held on the left or right arm of his nurse, and he would not take it in his left hand until a week later although I tried to make him do so; so that the right hand was a week in advance of the left. Yet this infant afterwards proved to be left-handed, the tendency being no doubt inherited—his grandfather, mother, and a brother having been or being left-handed. When between eighty and ninety days old, he drew all sorts of objects into his mouth, and in two or three weeks' time could do this with some skill; but he often first touched his nose with the object and then dragged it down into his mouth."

In this description we clearly see what importance C. Darwin attached to movements, and to movements in children as modes of expression of infantile development, especially as to the nerve-system and capacity for mentation.

Bibliography.

The following works are referred to in the text, or contain matter of interest in relation to the modes of expression :—

1519, Leonardo da Vinci died this year. His "Treatise on Painting" published 1651.

1644, Bulwer, John, "Chirologia, or the Natural Language of the Hand," etc.

1649, Bulwer, John, "Pathomystomia, or a Dissection of the Significative Muscles of the Affections of the Mind," etc.

1667, Le Brun, "Conférences sur l'Expression des différents Caractères des Passions."

1747, Parsons, James, M.D., "Human Physiognomy explained." Illustrated. This author gives a list of forty-one old authors who have written on Physiognomy. *Philosophical Transactions.*

1750, Mengs, Antony Raphael, the works of, translated 1796.

1775, Hartley, "Theory of the Human Mind." Edited by Joseph Priestley, F.R.S.

1777, Gregory, John, M.D., "A Comparative View of the State and Faculties of Man, with those of the Animal World." Seventh edition.

1788, Blaine, Sir Gilbert, Croonian Lecture.

1792, Camper, Pierre, "Discours sur le Moyen de représenter les diverses Passions," etc." Illustrated.

1805, Moreau (de la Sarthe).

1807, Lavater, G., "L'Art de connaître les Hommes." Illustrated.

1822, Siddons, Henry, "Practical Illustrations of Gesture and Actions," from a work on the subject by M. Engel, member of the Royal Academy of Berlin.

Sartandière, "Physiologie de l'Action musculaire appliquée aux Arts d'Imitation."

1844, Bell, Sir Charles, "Anatomy and Philosophy of Expression." Third edition. Illustrated.

1853, Little, W. J., M.D., "On Deformities." Illustrated.

1855, Spencer, Herbert, "Principles of Psychology."

1862, Duchenne, "Mécanisme de la Physionomie humaine."

1865, Gratiolet, "De la Physionomie."

1865, Lemoine, Albert, "De la Physionomie et de la Parole."

1868, Trousseau, "Clinical Medicine." New Syd. Soc. translation. In article on Tetany, describes the "convulsive hand."

1868, Tyndall, John, "Heat a Mode of Motion." Third edition.

1872, Darwin, Charles, "The Expression of the Emotions in Man and Animals." Illustrated.

1872, Tuke, Daniel Hack, "Illustrations of the Influence of the Mind on the Body."

1874, Marey, E. J., "Animal Mechanism."

1874, Meillet, H., "Des Déformations permanentes de la Main." Illustrated.

1876, Ferrier, David, F.R.S., "The Functions of the Brain."

1877, Charcot, "Lectures on the Diseases of the Nervous System." New Syd. Soc. translation. Describes the "writing hand" in Paralysis agitans—also other postures. Illustrated.

1878, Roberts, Charles, "A Manual of Anthropometry."

1879, Lindsay, N. Lauder, M.D., "Mind in the Lower Animals." With Bibliography.

1883, Romanes, George John, F.R.S., "Mental Evolution in Animals."

CHAPTER XIX.

METHODS AND APPARATUS FOR OBTAINING GRAPHIC RECORDS OF MOVEMENTS IN THE LIMBS, ETC., AND ENUMERATING SUCH MOVEMENTS AND THEIR COMBINATIONS; PROBLEMS TO BE INVESTIGATED BY THESE METHODS.

Movement as a result of vital action is capable of physical experimentation—Early attempts to record movements—Apparatus described: the motor gauntlet; the recording tambours; the contact-making tambour; electrical counter; method of using the apparatus—Problems; as to muscular twitching in exhaustion—Movements of an infant—Inhibition by light—Measurement of differentiation of movements—Retentiveness—Signs of emotion—Potentiality for mind—Co-ordination—Athetosis—Chorea.

THROUGHOUT this essay I have endeavoured, as far as possible, to describe modes of expression in terms of movements and results of movements. This has been done purposely, for it has long been my conviction that, of all the results of vital action, movement is the outcome or result most suitable for direct experimentation.

About five years ago I commenced some experiments with the object of producing a mechanical method of recording movements. At first I fixed writing points to the tips of the fingers, making them touch a travelling sensitive surface. This was a clumsy and inefficient plan. Various arrangements of tambours attached to the moving parts were tried, after the methods employed by M. Marey, and described in his work.

At length it occurred to me that an arrangement might be used attached to the hand, but leaving it free to move.

The following account was published in the *British Medical Journal*, September 22, 1883:—

The apparatus * employed consists of—

1. A "motor" made of india-rubber to be attached to the hand, one tube to each finger, or moving part (Fig. 46, A, B, C, etc.). From these, pieces of thin conducting-tube (a, b, c) carry air to a set of recording tambours.

2. A frame supporting the recording tambours, and electrical signals (Fig. 49).

3. A new "electrical contact-making tambour" (Fig. 50). It is a modification of the Marey tambour adapted to the purpose of actuating an electrical counter.

4. An electrical counter (Fig. 51).

Now, as to the further details of each piece of the apparatus and its uses. In the motor which is attached to the hand (Fig. 46), the principle employed is as follows:—When a cylindrical tube

---

* Towards the expenses of this apparatus, a grant was made by the British Medical Association on the recommendation of the Scientific Grants Committee.

closed at one end is bent, flattened, or compressed, its capacity is lessened, and therefore air is pressed out of it and driven into the tambour with which it is connected. The needle of the recording tambour is thus moved by the bending of the tube of

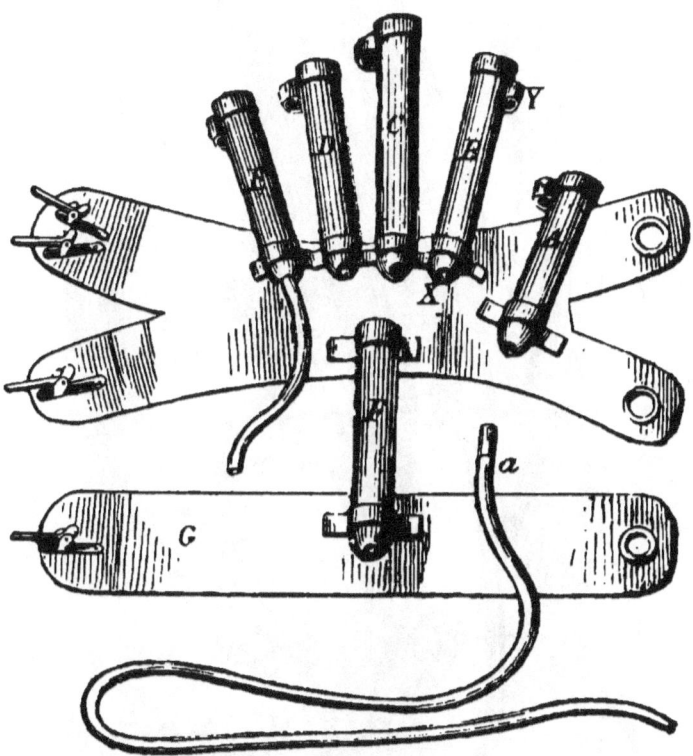

Fig. 46.—Motor Gauntlet.

the motor attached to the finger; the movement of the recording needle corresponds to the movement of the finger and the activity of that part of the central nerve-mechanism which moves the finger. These tubes are moulded for me in soft red rubber by Messrs. Warne and Co.; they are 7¾

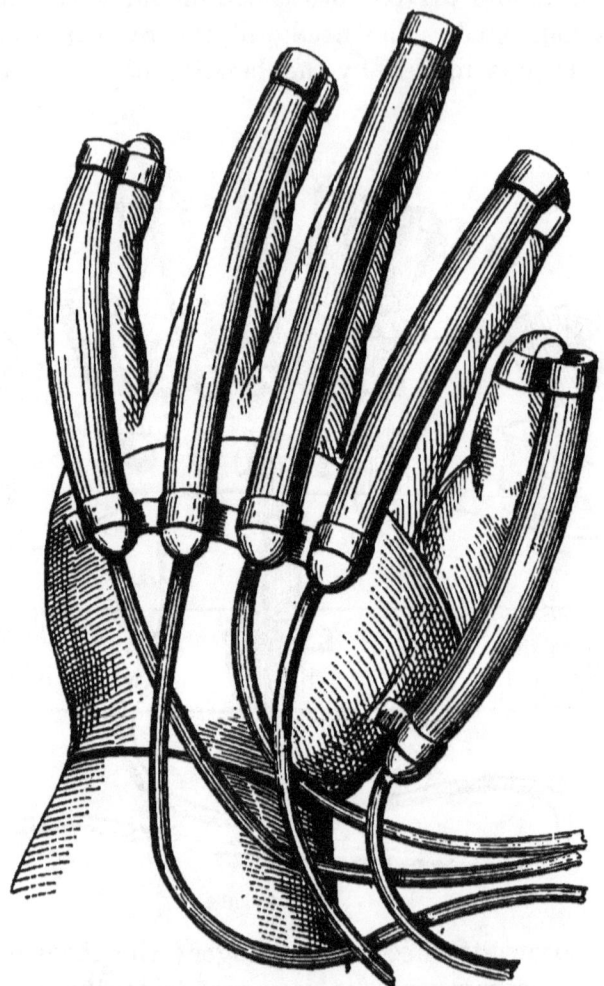

Fig. 47.—Motor Gauntlet on Hand.

millimètres in external diameter; one end of the tube, X, is pointed and pierced, the other end, Y, is closed by a moulded cap (Fig. 46, *Y*); and a band moulded to the cap gives the means of fastening the tube to the finger. These tubes are mounted on a "foundation" of sheet red rubber, which is fastened round the hand like a glove (see Fig. 47). The arrangement represented in Fig. 48, *a* and *b*, is convenient in the case of adults, in whom the thickness of the finger makes a considerable drag upon the upper surface of the tube during flexion of the finger. It consists of a pair of moulded caps mounted on a ring back to back: the septum between the two is divided halfway down, and the lower portion is pierced, allowing air to pass from

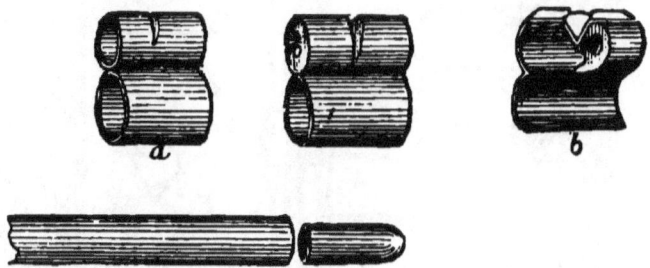

Fig. 48.—Junctions for Motor Tubes.

one cap to the other. The band is fastened to the middle of a phalanx, and, the finger-tube being cut at this point, the divided ends are inserted one in each cap. The whole is thus rendered air-tight, and the tension of the upper surface of the tube during flexion of the finger is thus relieved.

The recording tambours mounted in a frame (see Fig. 49) give tracings upon a revolving cylinder, showing the frequency and coincidence of the various movements. The electrical signals give the means of marking time, indicating events, etc.

"The electrical contact-making tambour" is represented in Fig. 50. Across the ordinary shell of a Marcy's tambour, a main bracket, *a, a*, is fixed, to which three smaller ones are screwed, having holes in them to take two arbors. One

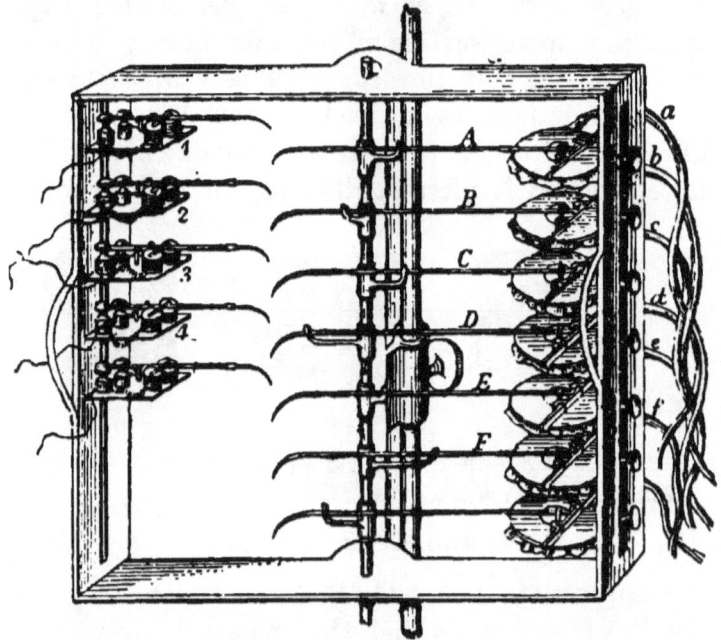

Fig. 49.—Frame supporting the recording tambours, and electrical signals.

arbor, *b*, has a beam, *d, e*, with a small weight at one end, *e*, and is connected by a short link with the india-rubber head of the tambour. To the other arbor, *C*, a block of ebonite is fixed to insulate two wires, *f, g*, with platinum ends, which lie under and nearly touch the beam *d, e*. Each of these

CONTACT-MAKING TAMBOUR. 353

wires is connected with terminals *h* and *i* by a coil of thin copper wire. A light bent blade-spring is inserted between the ebonite block and the bracket,

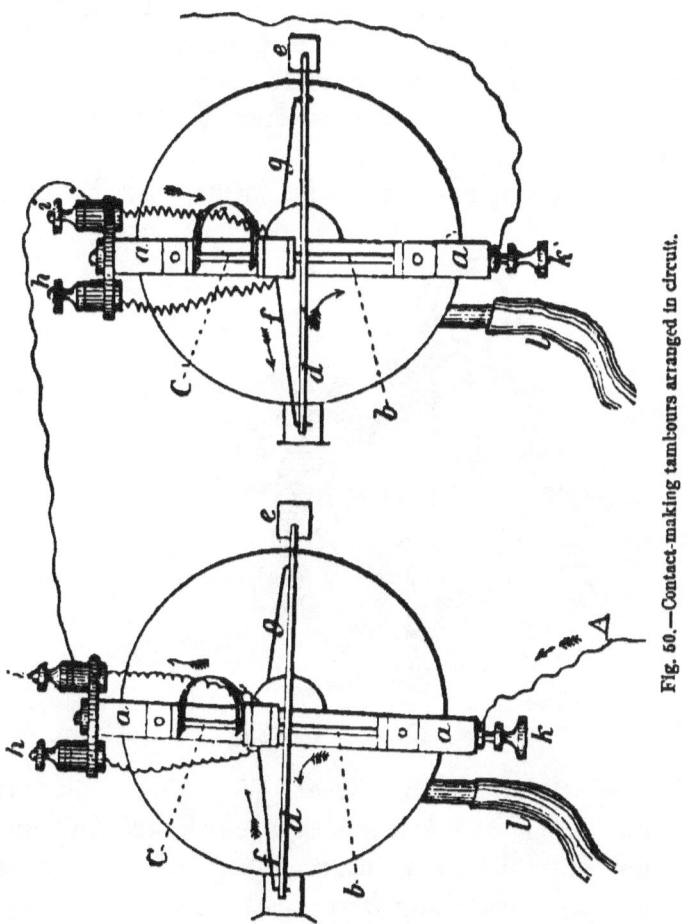

Fig. 50.—Contact-making tambours arranged in circuit.

to resist the motion of the arbor *C*. The beam *d, e* has platinum soldered on the part where the platinum ends of the wires are to touch. A terminal, *k*, is fixed to the main bracket. Increase of air-pressure in the tambour raises one end of the beam, *e*, and depresses the other, *d*, which thus comes in contact with its insulated wire *f*, completing a circuit through the terminal *i*; diminished air-pressure allows the beam to move in the opposite direction, completing a circuit through the terminal *h*.

The instruments are made for me by Mr. W.

Fig. 51.—Electrical Counter.

Groves, 89, Bolsover Street, W. The electrical counter (Fig. 51) consists of a clock, to the pendulum of which a small piece of soft iron, *c*, is attached, opposite an electro-magnet, *d*. In use for counting the movements of a finger, the conducting tube (Fig. 50, *l*) from the finger is attached to the electrical contact-making tambour. One pole of a Leclanché battery of four cells is connected with k, and a wire connects i with a in Fig. 51, while b is connected with the other pole of the battery. Each flexion of the finger thus allows the escapement of the clock to pass one tooth, and the hands indicate on the dial the number of finger-flexions, and the number of times the portion of the nerve-mechanism connected with the finger-flexion has been in activity.

If it be desired to record by figures the number of times that certain special movements (coincidences of movement) have been made, we may proceed as follows:—Let us consider only flexor movements of the fingers. There are five digits; it is required to count how many times each of the possible combinations of flexion of the five digits occurs. There are thirty-two such possible combinations of flexion. Let the digits be called A, B, C, D, E; there are thirty-two possible combinations of these five factors, viz. ABCDE, AB, ABC, ABCDE, etc. Now, by the following method, it is possible to show, on thirty-two counters, how many times each of these possible combinations of finger-flexion has occurred in a given length of time.

Say it is desired to make a counter enumerate how many times the digits A and B have flexed together. The conducting tubes from the fingers A, B are to be connected each with a contact-tambour; then, by making the electrical connection between these as indicated in Fig. 50, and including the counter in the circuit, that counter will be actuated every time that the fingers A, B bend together, and will enumerate how many times this special combination of movements has occurred.

This apparatus has been used, and results have been obtained, but the labour of obtaining results is very great. During the last year I have been endeavouring to make the apparatus more complete and more serviceable. I am replacing all electrical apparatus by pneumatic mechanism, as being simpler; numerous causes of delay have arisen, but when the whole is completed a full account will be published. Experiments are arranged for recording the movements of the eyes, head, and fontanelle, but they are not completed. As to the uses of this means of investigating the nerve-centres, I will state the propositions and problems, and say how I propose to follow out such inquiries.

As to the problems, and how to work at them.

In each case, for the sake of simplicity, we will suppose hand movements, and the nerve-centres causing them, to be the subjects of inquiry.

(1) In exhaustion, finger twitching is sometimes an expression of the condition. Here the number of finger twitchings in a given period of time, say one hour, may be counted; we can enumerate the frequency with which each separate finger has moved, and also the frequency with which any combination of fingers twitched together. The external circumstances may be varied; say the enumerations are taken for half an hour while the subject is alone and unoccupied, then during half an hour while he is engaged in con-

versation. We can thus obtain some information as to the effect of conversation upon the nerve-centres of an exhausted man. Conversation may excite or inhibit the separate discharges of the nerve-centres, and may cause them to occur in larger or smaller combinations; experiment alone can demonstrate these points.

(2) The amount of spontaneous finger movements in an infant may be similarly demonstrated by enumeration and by graphic records.

(3) The effects of the sight and sound of objects upon the action of the nerve-centres, and particularly the effect in producing special combinations of movements, will be shown best by taking tracings of those movements. The time of the sight of an object, or of hearing a sound, is easily indicated on the tracing by the action of an electrical signal. The inhibitory action of light upon the spontaneous movements of an infant is shown by a tracing (see Fig. 9, p. 101).

(4) Applying motor tubes over all the joints of the upper extremity, and enumerating the movements, will show (*a*) the relative frequency of movement of large parts as compared with small joints; (*b*) if tracings be taken, the frequency of interdifferentiation of movements will be demonstrated; that is, the frequency with which fingers move without the elbow, or *vice versâ*, etc.

(5) Retentiveness, or permanent impressionability, is best demonstrated by the graphic method. If the same stimulus on different occasions produces similar movements, we may conclude that the

nerve-mechanism is retentive, and is not very changeful.

(6) Special combinations of movements, as the outcome of certain brain conditions or emotions, may be demonstrated by enumeration or by the graphic method. A series of movements would best be demonstrated by the graphic method. Thus, I think that some of the signs of general brain conditions may be recorded by numbers, with as much accuracy as the temperature of the body, which is one sign of fever, and can be recorded by the thermometer.

(7) All those signs in an infant which show its capacity for the development of mind, in as far as mind is expressed in movements, might be recorded by the graphic method in terms of lines, and enumerated in figures corresponding to isolated movements and combinations of movements. By repeating the examinations at different ages and under various external circumstances, expressions may be obtained, in terms of lines and figures, indicating the evolution of the physical indications of mind. At the same time, it would be easy to observe the effects of external circumstances upon this evolution.

(8) A very interesting question is to find out how co-ordination of movements is brought about. A co-ordinated movement is a series of movements, a series of combinations of movements, and single movements. A series of movements can be recorded by the graphic method; the time of each movement is indicated by the tracing, and the time of the

action of external forces, such as seeing a light, may be marked on the same paper by one of the electrical signals.

Here, then, there seems to be data for observations as to the production of co-ordination by external forces.

To give an example in detail. Take a healthy active infant, six months old. Adapt the motor gauntlet to its hand, and make connections with the other part of the apparatus. Take tracings of the movements, and enumerate them and their combinations.

Now show the child an orange; let the light reflected by the orange fall well within its field of vision. Note the time of showing the orange upon the tracing, by means of one of the electrical signals. The co-ordinated movements of the infant's fingers, wrist, and elbow are indicated by the recording points. If all the fingers flex together the combination of that special set of movements will be recorded by the enumerator for that special combination of finger movements. Repeating the experiment several times, we shall find out whether this special combination of movements occurs more frequently and certainly upon the stimulus of being shown the orange, than it does in simple spontaneous movements of the fingers. A nurse would say the child's attention is attracted by the orange when it sees it. If after many such trials on many days the special movements of the hands occur with greater certainty and quickness on seeing the orange, the nurse would say that the infant has learnt to know or recognize the orange, that it remembers the orange,—and so it is.

The growing power of co-ordination of movements is expressed by the enumeration of certain movements counting a higher figure during that period of time in which the infant has often been shown the orange.

Enumeration will show the increasing co-ordination; tracings will show the interval of time between the stimulation and the co-ordination of movements resulting.

If instead of making observations on a healthy child, with a healthy brain, the experiment be made on one whose brain was defective at birth, we may get very different results. A special feature of the spontaneous movements of an idiot consists in the repetition of a very few combinations, occurring over and over again, in monotonous and useless succession. Such movements are seen in the condition termed athetosis (see chap. vi.).

By saying that these combinations of movements are useless, I mean they are not excited by external stimuli directly, and are not in unison with, and do not vary with the external stimuli, and so they are useless to the individual. Such a child, owing to its defective brain, would give results on our enumerators showing that certain combinations only, occur with a high degree of frequency. Otherwise expressed, we may say that the high degree of frequency of a very few combinations of movements, together with the fact that these combinations are not readily changeable by the action of external stimuli, indicate the brain defect.

In studing chorea by these methods, it has been possible to show that cases differ from one another in the character of the movements; in some cases the movements are singularly simple, in others each gross movement is compounded of several smaller ones. Again, it has been possible to show what causes increase or diminish the amount of movement in a limb; the effects of talking to the patient, keeping her in darkness, making her sit up in bed, etc., in as far as these circumstances affect movement can be investigated. Tracings have been taken showing ankle-clonus; senile tremor; movements of the sternum in health and in chorea; the movements of athetosis, and those due to the sucking of an infant, etc.; the movements of the limbs due to pulsation.

I trust that in this chapter it has been shown that there is reason for trying to describe brain conditions in terms of movements and the results of movements, inasmuch as these signs are capable of observation and experimental inquiry. I cannot tell what results may come from such work, but, as previously said, it was undertaken with the desire to determine what forces are necessary to aid brain development, such as is necessary to the mental and moral evolution of the child.

INDEX.

A

Absence of movements expressive, 227, 263
Action described as a series of movements, 70
—— kinetic, 45
—— trophic, 33
Activity, expression of, 233
Actor on stage, 40
Afferent force eliciting expression, 37, 42
Amœba described, 83
Analogy between chorea and movements in plants, 111
—— between series of trophic and kinetic actions, 278
Analysis of expression, 255, 257
—— of movements, 75
Anatomical analysis of movements, 79
—— —— principle of, 75
—— description of a movement, 75
Anatomy of head movements, 183
—— of the upper extremity, 155
Anger, expression of, Bain on, 50
—— Bell's description, 51
—— Siddons' description, 51
Antecedent force is indicated by every movement, 49
Antithesis, principle of, 319; referred to by Bulwer, 325
Ants lose their wings, 39
Anxiety, expression of, 197
Apparatus for a reflex action, 36
Army corps, its movements, 74
Art, average productions, 317
—— may teach physiology, 289
Artistic composition, 313, 317
—— expression by postures, 152
—— touch, 293
Ascidian, its nerve-mechanism, 84
Aspect of anxiety indicates a cause of fear, 16
—— of headache, 206
Assumption that mind depends upon structure, 8
Asymmetry of action, 78
—— of ears, 274
—— of face, 200
—— of movement, 78
Athetosis, 127
—— of face, 208
Attention attracted, 59, 249, 251
Attributes of a property, 267
—— of a function as modes of expression, 277

B

Bain, Professor, on anger, 50
Barrett, W. F., on the sensitive flame, 17
Beard, its growth at manhood, 38
Beauty, 313, 314
—— emotion of, 237
—— not to be represented by weakness, 314
Beautiful faces, 313
Bee, movements of, from flower to flower, 62
Bell, Sir Charles, description of anger, 51; of joy, 258; of the Dying Gladiator, 303
—— on laughter, 52, 258; on the permanent form of head in contrast to expression, 182

INDEX

Bibliography, 344
Biography of an infant, 342
Biology capable of advancement by mathematical processes, 281
Blane, Sir Gilbert, 332
Blindness, congenital, 274
Body, proportions of, 278
Brain, areas affected in chorea, 115
—— disease may produce special postures, 169
—— evolution of, in child, 283
—— of monkey, experiments on, 86
—— properties necessary to mentation, 242
Bulwer, John, on art, 281; on movements, 321

C

Cain, statue of, 301, 319
Camper, 331
—— his value of metaphysical reasoning, 305
—— on expression of joy, 307
Capacity to feel joy, 42
Cases of athetosis, 127
—— of coincident defective development, 133
—— of expression from brain disease, 208
Cast of face, 21
Cave animals lose sight, 101
Cerebral facial palsy, 107
—— localization, 108
Change of function, 38, 39; in stomach, 39
—— of postures, 143
Character, judgment of, 15
Chorea, 110, 115, 361
Child at play, 282
—— its impressionability, 20
—— study of a, 284
Children growing like parents, 285
—— irritable, 280
—— nervous, 117
Circumnutation in plants, 26, 280
City life, 62
Classification of movements, 79
—— of spontaneous postures, 151
Cleft-palate, 136, 275
Chlorophyll, significance of, 22
Coincidence of action of many or of few members or units, 284
Coincident development of parts, 42, 273

Coincident development of postures, 151; of head and hand, 151, 188
—— similar development, 273
Collateral deviation of head and eyes, 105
—— differentiation, 77, 179
Colour a means of expression, 27, 38
—— contrasted with form, 317
Coloured vision with headaches, 121
Combinations and sequences of action, 279
—— of movements, 70
Combined movements, Bain on, 235
Comparison of a kinetic and a trophic series of actions, 277, 281
—— of an idiot and an intelligent man, 13
Complacency, Camper's description of facial expression, 306
Composition of figure-drawings, 313-317
Compound movements, 112
Conflict of muscles in face expressing emotion, 213
Congenital blindness, 274
Consciousness, 225
—— its physical indications in man, 226, 262
—— not the cause of physical phenomena, 189, 226
Contempt, expression of, described by Camper, 307
Co-nutrition in two subjects produces similar proportions of growth, 286; and similar series of movements, 286
Convulsive hand described, 156
Co-ordinated movements, 71; of eyes, 227
Co-ordination by external forces, 286
Corpus striatum, destruction of, 105; irritation of, 106
Criteria of life, 12
—— of mind, 28, 252
Criterion of a property, 12
Crystals, growth of, in process of repair, 40, 41

D

Dance, movements of, 80
Darwin, C., biography of an infant, 342
—— description of laughter, 257
—— on snarling, 15

INDEX. 365

Darwin, C., on stamping of rats, 54
Deaf mutes, 274
Defiance, 293
Deformity to be avoided in art, 314
Delayed expression, 250
Depressing news, effect on face, 201
Description by means of a series of kinetic or trophic actions, 278
—— in anatomical terms, 178
—— of a germinating seed, 279; analyzed, 280
—— of growth and movement compared, 281
—— of the growth of a frog, 281
Despair, 309
Development, a mode of expression, 35
—— indicated by growth and movement, 71, 278
—— of an infant, 247
Developmental defects, 133
Diana, statue of, 143, 298
Diathesis, 275
Difference between living and non-living things, 268
Differentiation of ovule, 23
Direct expression by postures, 148
—— —— defined, 20
Disease, expression of, to be avoided by artists, 294
Drawing of modern figures, 190
Drooping wrist, 168
Drosera rotundifolia, 152
Dull countenance, 273
Dying Gladiator, 145; Sir C. Bell's criticism, 303

E

Ears, asymmetry of, 274
—— deformity of, 134, 137
Education, defects of, 313
Emaciation, 275
Emotion evidenced by postures, 149
—— expression of, 43
—— of the beautiful, 237
Emotions, 234
—— excite movement, 312
—— how to be studied, 226
Empirical expression defined, 21
Energetic hand, 164
—— speaker, his postures, 143
Environment, or external force, its importance, 282, 285

Epicanthic fold of eye, abnormal, 137, 273
Epilepsy, study of, 110
Epinasty, a mode of vegetal growth 39
Equal proportional growth, 271
Equilibrium of posture, 58
Evolution of child's brain, 283
Excitation of weak nerve-centres (a principle), 179
Exhaustion, signs of, in man, 229
Experimental method, 248
Explanation of the nervous hand, 167
Expression by a series of movements and trophic actions, 279
—— by change of function, 38
—— by colour, 27, 38
—— by form, 27
—— by movement, 39, 48
—— by results of movement, 53
—— by sound, 38
—— by temperature, 27
—— described by Camper, 305
—— direct, 20
—— empirical, 21
—— free and mobile, 182
—— in the eye, 217
—— of feeling, 292
—— of the emotions, 43
—— the term defined, 11
External forces, importance of their study, 282
—— —— regulate trophic and kinetic series, 282
Evolution of the brain, 283
—— resulting from the environment, 285
Eyeball distinguished from the eyelids, 214
—— nearly expressionless, 217
Eyelids, 217
—— movements of, Dr. Gowers on, 223
Eye-movements during sleep, and coma, 227
—— intellectuality of, 222, 224
—— express condition of the mind, 221
Eyes, application of principle of small and large parts, 224
—— attracted, 144
—— attracted and repelled, 220
—— attraction and repulsion of, under visual stimulus, 222
—— engaged, 222
—— free or disengaged, 222

366 INDEX.

Eyes, horizontal movements of, 222; compared with the vertical, 223
—— lateral movements of, 219
—— loss of associated movements of, 218; in sleep, 227
—— lost from want of light in cave-living animals, 101
—— movements of, in a train, 220
—— spontaneous movements of, 222; in infant, 220
—— upward movements compared with the vertical as to their intellectuality, 219

F

Face, an index of the mind, 193
—— action of gravity on, 147, 202
—— anatomy of, 194
—— asymmetrical, 200
—— athetosis of, 208
—— defined, 194
—— dull or bright, 198
—— engaged, and not free to express emotion, 207
—— fatness of, 274
—— how to examine, 196
—— in fatigue, 229
—— in headache, 119
—— intellectual or vulgar, 198
—— its muscles and nerves, 194
—— long, 201, 229; seen in fatigue, 205
—— method of analyzing, 196
—— movements of, 195; in a case of brain disease, 208, 211
—— nutrition of, 198
—— usually free, 144
Faces of idiots, 198, 204
Facial expression: of mental suffering, 197; of bodily suffering, 200; of fatigue, 205
—— paralysis, 202; from brain disease, 107
—— signs of previous nerve-muscular action, 199
—— zones defined, 196
Fat, absorption of, in face, 198
Fatigue expressed in the face, 105
—— signs of, 58, 228
Features, handsome, 274; regular, 274
—— of the face, 141, 197
Fear, 253
Feeble hand, 158

Feeling expressed by nerve-muscular action, 292
Ferrier's experiments on the brains of animals, 90
Figure-drawing, 317
Finger-twitching, 125, 229
Fist, clenched, 57
Flexion and extension of head, 184
Formula giving the possible number of combinations of a given number of units, $2n$, 70
Foster, Professor, on mechanism of voice, 54
Free, or disengaged parts, 144, 315
Frequency of movements, 69
Fullness under the eye, 120
Function in two subjects compared as to time and quantity, 269

G

Gait in walking is a series of movements, 72
—— reeling, 293
—— spiritless, 293
Gang of slaves working in unison, 74
General excitement of nerve-centres, 179
—— weakness of nerve-centres, 180
Geranium with pink flowers of feeble constitution, 22
Germination of seed, 279
Glands change their function, 39
Good proportional growth, 277
Graphic method, 348
Gravity, action of, on face, 147
—— on living beings, 146
—— effect in producing postures, 146
—— of, on head, 147, 188, 192
Greek vases, 299
Gregory's works, 327
Grinning, 203
Growth, 32
—— a series of trophic actions, 279
—— indicates life, 33
—— is a material change, 33
—— of body, 41
—— results from nutrition, 33

H

Hair, colour of, 42
—— grey, 41
—— on face, 39

INDEX.

Hand, an index of the mind, 295
—— convulsive, 156
—— drawn away from a hot object, 36, 253
—— energetic, 164
—— feeble, 158
—— free or disengaged, 316
—— in fright, 157
—— in rest, 159
—— nervous, 163
—— often not expressive in art representations, 317
—— straight, 160
—— straight with thumb drooped, 161
Hands gone mad, 129
Handsome features, 274
Hartley's theory of the human mind, 325
Head, application of principle of contrast of small and large parts, 189
—— flexion and extension, 184
—— inclination, 184
—— its axes described, 183
—— its postures and movements defined, 184
—— measurements, 243
—— method of describing movements, 184
—— movements effected by light, 98
—— not free or disengaged, 144
—— rotation, 184
—— shape of, in idiots, Dr. Shuttleworth on, 191
—— usually free, 144, 189
—— weak posture of, 188
Headaches, facial appearance of, 120
—— in children, causation, 128
Healthy activity, 233
Heart, physical signs of its action, 14
Heat expressed by temperature, 14
—— how studied by Tyndall, 340
Hemiplegia described, 105
"Hercules at Rest," statue, 304
Heredity, 43, 46, 285
—— physical properties its necessary accompaniments, 46
Horror, expression of, 151
Hydrocephalus with athetosis, 199
Hyper-extension of knuckles, 166
Hyponasty, a mode of vegetable growth, 39

I

Icthyosis, 138, 274
Idiot but slightly sensitive to pain, 253
Idiot compared with an intelligent man, 13, 242, 277, 278
—— facial expression of, 204
—— microcephalic, 274
Idiots' faces, Dr. Langdon Down on, 198
Illness may prevent expression, 59
Ill proportion of the body, its significance, 278
Imitation of anger, 40
Impressionability, 32
—— absent in young infant, 244
—— consideration of its attribute "time," 251
—— in nerve-mechanism, 233
—— its expression delayed, 228
—— permanent, 33
Impression of past nerve-muscular action in face, 199
Inability in the infant to move the hand straight, 244
Inclination of head, 184
Inco-ordinated movements, 71
Index of the mind, 193
—— of the hand, 295
Infant at birth described, 96, 243
—— biographical sketch of, by C. Darwin, 342
—— compared with the adult, 242
—— mindless at birth, 243
—— nerve-system of, 293
—— table of its development, 250
Infants ill proportioned, 278
Inhibition, 94
—— by light and sound, 100, 101
Instinct, 234
Intellectual face, 198
Intellectuality of action in each facial zone, 203
—— of eye-movements, 222, 224
Intelligent and non-intelligent movements, 79
Interdifferentiation of condition of parts, as a principle, 76
—— of postures, 178
Iris nerve-supply, 215
Irritability, expression of, 229
Irritative lesion of brain, 106
Iron may be magnetized, 39

J

Jaw, movements of, 192
Joy, described by Bell, 258
—— expression of, Camper on, 307

K

"Kind," as an attribute of a property, 268
Kinds of movement, 268
—— of movements in an infant, 248
—— of property, described in terms of time and quantity, 8
Kinetic action, 45

L

Labourer digging, 144
Langdon Down, Dr., on idiots' faces, 198
"Laocoon," Camper's criticism of, 304
Large heads in children, 278
Larynx, changes occurring in, 39
Lateral deviation of head and eyes in hemiplegia, 105
Laughing humour, 16
Laughter, Bell's description, 53
—— Darwin's description, 257
Laws of nature, 8
Leaves, ill developed, 274
Left hand weaker than the right, 126
Life, signs of, 12
—— non-understandable, 12
Light, action of, on brain, indicated by reflex action, 98-100
—— —— on vegetation, 103
—— inhibitory action, 100
—— its action on the eye, 96
—— movement of, attracts attention, 99
—— produces permanent impressions, 99
—— stimulates head rotation, 98
—— trophic action of, 101
Localization in the brain, 108
Locomotion, 73
Locomotor ataxy, gait of, 72
Long face, 201
Ludicrous idea, 53
Lungs of fœtus, 39

M

Magnet shows change of function, 39
Man, study of, 294, 295
Manhood, signs of, 38
Marasmus, 276
Marey, M., on walking, 73
Marshall Hall's works, 537

Materialistic views in science necessary, 240
Measurement of head and body, 277
Mechanism of a reflex movement, 43
—— of the eyelids, 217, 223
—— of postures, 141
Memory, 246
—— indicated by movements, 246
Mengs, Antony Raphael, on painting, 310
Mental anxiety, expression of, 197
—— states, expression of, 295
Mentation, the term defined, 5
Metaphysical abstractions not considered in this work, 67, 226
Method of analyzing a face, 196
—— of biological description, 281
—— of demonstrating a posture to be nerve-muscular action, 168
—— of physical inquiry as to mind, 241
Microcephalic idiot, 274
Mimosa pudica, 24; experiments upon, 113
Mind in the infant, criteria of, 244
—— physical study of, 28, 241
—— what is it? 252
Mobile expression, 183
Modern pictures, 316
Modes of expression, 12
Molecular motion, probably the cause of coincident development, 67
Motion, physical investigation of, 6
Movement a mode of expression, 39, 48
—— a visible and physical action, 48
—— capable of measurement by graphic method, 67
—— complex in the adult, 61
—— correlatable with force, 49
—— in art indicated by postures, 152
—— indicates antecedent force, 49
—— in one subject, 58; in two or more, 68
—— in the body is change of posture, 142
—— its time and quantity, 68
—— kinds of, 268
—— lessened by bad feeding or illness, 231
—— manner or kind of, 48
—— notable by two observers, 49
—— of an aggregation of members, 74
—— of the whole subject or of its parts, 74

Movement, subsidence of, 58
Movements, analysis of, 75
—— classified according to attributes, 67; according to mode of production, 67
—— combinations of, 68
—— co-ordinated, 71
—— described in anatomical terms, 75
—— in aggregation or in succession, Bain on, 236
—— in children, 244
—— inco-ordinated, 71
—— in plants, 21, 114
—— in regular series, 80
—— intelligent and non-intelligent, 79
—— of a bee from flower to flower, 62
—— of an actor in anger, 36, 40
—— of hand described in terms of postures, 281
—— said to proceed from feelings, 80
—— spontaneous, reflex, voluntary, 60, 67
—— stimulated by external forces, 100, 247
—— two or more considered in relation to time, 68
Musical box attracts child's attention, 20
Music, soothing effect of, 233
Mutes born deaf, 274

N

National modes of expression, 266
Negation, expressed by head rotation, 192
Nerve-centre, 91, 168
Nerve-mechanism of vertebrates, 84
—— of Ascidians, 83
—— of reflex movement, 36
Nerve-muscular action, 85
—— —— in Art, 107
—— —— may express feeling, 292
—— —— study of, necessary to Art, 294
Nerve-supply to face, 194
Nerve-system, its importance, 1
Nervous children, 117
—— hand, 163, 297, 299
Newton, Sir Isaac, 327
Nutrition, an expression of life, 33
—— expressed by movement, 59
—— expression of, 230

Nutrition, its signs, 23, 231
—— its total outcome in trophic and kinetic function, 283
—— of vegetable ovule, 23
—— produces trophic and kinetic functions, 283
—— —— growth and movement, 283
—— the term defined, 22

O

Objective, or physical facts, 12
Observation on material structure, 6
—— of movements as well as histology wanted, 6
—— on ingoings and outcomings, 7
Organic postures, 146
Orbicular muscles of eye in sleep, 227; relaxed in headache, 120, 206
Outcome of a subject always expressive, 14
—— of any function an expression, 38
Oxalis, its pulvinus, 25
—— *corniculata*, imperfect pulvinus, 26

P

Pace defined as a series of movements, 73
Paget, Sir James, on repair of crystals, 40
Pain, 236, 253
—— not itself a cause of movement or expression, 17
Passionate children, 117
Patent foramen ovale, 275
Pea plant grown in darkness, 232
Permanent impressionability, 33
—— —— may be indicated by reflex action, 33
—— —— opposed to evolution, 83
Phlegmatic temperament, 276
Phonograph, mechanical action necessary to its working, 19
—— shows permanent localized impressionability, 19
Physical forces aid development of mind, 4
—— pain, 200
—— phenomena capable of analysis, 3
—— signs of hidden properties, 4
—— signs of mind, 240

Physiognomy, 2
—— its signs unsatisfactory, 2
—— of scrofula, 275
Physiological principles of analysis, 77
—— study useful to art, 294
Pink geranium of feeble constitution, 76
Plants, growth of, 285
—— movements of, 26
—— unicellular, 23
Plaster cast of head, 21
Playfulness, 249
Political procession, its significance, 74
Pope's description of Achilles, 260
Porcupines, noise made by, 55
Postulate as to nerve-centres, 168
—— physical phenomena due to physical causes, 2
Posture, term defined, 57, 140
Postures a means of describing movements, 281
—— classification of, 151
—— difficulty of description, 162
—— due to brain disease, 169; to reflex action, 148
—— fallacies concerning, 149
—— in animals, 152
—— in art, 152
—— in plants, 152
—— organic, 146
—— the result of last movements, 57
—— voluntary, 150
Potentiality for mind, 242
Pottery presents ancient figure-drawing, 299
Primitive combined movements, Bain on, 235
Primary meristem, 23
Principle as to small and large parts (interdifferentiation), 76; applied to head, 189; applied to eyes, 224
—— contrasting large parts and small parts as to posture or movement, 76
—— —— collateral parts (collateral differentiation), 77
—— of antithesis, 181
—— of general excitement of nerve-centres, 179
Principles of analysis and composition (artistic), 109, 317
—— —— of movements, 75
—— —— of postures, 178

Principles of physiological analysis, 77
Problems as to coincidences of movement, 69
Properties demonstrated by external forces, 44
Proportional growth, 270
—— development a sign of nutrition, 232
Proportion of height, weight, age, 41, 279
Proportions of the human body, 107, 278, 311
Prostration of the body, 55
Pulse-tracing, 14
Pulvinus, 25
Pupil, 119, 215
—— in emotional states, 221
—— in sleep, 227
Puppy, its development described in terms of growth and movement, 71

Q

Quantity as an attribute of a property, 268
—— of growth, 269
—— of movement, 68
Quiet frame of mind, 201

R

Rabbits stamping, 54
Radicle of a seedling, its circumnutation, 280
Rate of movements, 68, 69
Records of movement capable of analysis, 49
Reflected action, 36
Reflex action, 92; congenital, 36, 244; acquired, 36
—— —— a mode of expression, 35, 36
—— —— at birth, 244
—— movements, mechanism of, 36, 83
—— postures, 148
Relation of the outcome to afferent stimulus, 37
Rest, a condition of nutrition, 232
—— expression of, 232
—— posture of hand, 159
Result of movements, 54
Retentiveness, 22
—— in brain, 234

Retentiveness indicated by reflex actions, 234
—— in infant, 246
—— not necessarily permanent, 34
—— to light, as in nerve-mechanism, 100
Root of plant, its mode of growth, 26
Rotation of head, 105, 184, 185, 187
Rules for analysis of expression, 256

S

Sanguine type, 275
School inspection, 262
Secondary movements, 55
Seed, growth of, 279
Sensitive flame, 17
Sensory centres in brain, 91
Sequences of action, 276 ; determined by external forces, 282, 284 ; not determinable by calculation, 70
Series of kinetic actions, 276
—— of movements, 276
—— —— and trophic actions, 279
—— —— as signs of evolution, 283
—— or succession of movements, 276; example of, 279
Shaking the mane, 55
Shame, indicated by head drooping, 293
Shuttleworth, Dr., on heads of idiots, 191
Siddons, H., on expression of anger, 52
Sight of a toy, its effects, 15
Signs due to an afferent force, 14
Similar growth due to an impression on the germ, 285
—— series of kinetic actions, 276
Similarity of development, 274
Sleep, 58, 227
—— disturbed, in children, 128
—— expression of, 189, 227
—— may prevent expression, 15
Small-headed children, 274
Small parts compared with large parts, as a principle, 76
Sneering, 201
Snow plant, 28
Sorrow, expression of, Camper on, 309
Sound, a means of expression, 38
Speech, a criterion of life, 13
Speed of movement, 68
Spinal muscles affected in chorea, 116
Spine bent in fatigue, 229

Spiritless gait, 293
Spontaneous and voluntary movements, 60
—— movement in children, 29
—— movements, 59 ; in infancy, 59, 244 ; lost in illness, 59
—— —— subside in deep sleep, 58, 65
—— postures, 148
Stages of growth, 28
Statue, expression of, 36
Stimuli, extrinsic, 101
—— mediate, 101
—— immediate, 101
Stomach, changes in its secretion, 39
Stooping attitude, 293
Straight hand, 160
Study of a child, 284
—— of a man, 294
—— of a man's face, 199
—— of a nervous subject, 261
Subject displaying a function, 280
Subjective conditions studied by their expression, 253
Subsidence of movement, 59, 61; in sleep, fatigue, and when attention is attracted, 58
Subsultus tendinum, 167
Summary, chap. ii., 11 ; chap. iii., 31 ; chap. iv., 61 ; chap. v., 80 ; chap. viii., 152 ; chap. x., 191 ; chap. xvi., 286
Surprise, expression of, Camper on, 307
Symmetry of movement, 77 ; its significance, 78
—— of postures, 179
Synchronous movements, 68, 69

T

Table of signs of development in an infant, 250
Teeth indicate age, 41
—— shown in anger, 54
Telephone exhibits impressionability, 18
Temperature indicates heat, 14
Thermometer indicates heat, 16
Thought, 254
Time as an attribute of a property, 267
—— occupied by reflex movement, 93
—— of action in two subjects, 269
Tooth-grinding, 119, 122
Tossing of head, 55

Tracing of a movement analyzed, 67
Treves, F., on physiognomy of scrofula, 275
Trophic action as an accompaniment of movement, 281
—— —— defined, 33; illustrated by growth of crystals, 40, 41
—— —— -expressive, 33, 40
—— —— of light, 101
—— and kinetic action abnormal in an idiot, 277
Turgid face in anger, 51
Twins as examples of coincident development or growth, 269, 285
Two or more movements considered in relation to time, 69
Tyndall, 341
Types of faces, 197

U

Unconsciousness, 227
Unequal growth in parts of plants, 39
Unicellular plant, 23
Uniform coexistence, a mode of expression, 12
Uniformities in nature, 8
Upper extremity, postures of, 154, 156

V

Vegetable cell growth, 26
—— life, reasons for reference to, 5

Venus de' Medici, 162, 296, 319
Vine branch, growth in darkness, 282
Visual impression, effects of, 99
—— perception, its effects, 15
Vital phenomena can only be studied by their physical signs, 28
Voice in illness, 231
—— in irritability, 230
—— mechanism of, 54
—— result of movement, 40
Voluntary movement, 60
—— postures, 150
Vulgar face, 196

W

Walking, 72
—— a series of movements, 235
—— said to be due to instinct, 235
Wanton look, Camper on, 309
Watch, movement of, 53
Weakness not a representation of high beauty, 214
Weak posture of head, 188
Weeping, 341
Winking, 200, 309
Wonder, expression of, 306
Work done, 56; by movement, 56
—— of art an expression, 56, 79

Z

Zones of face, 196

THE END.

www.ingramcontent.com/pod-product-compliance
Lightning Source LLC
Chambersburg PA
CBHW032010220426
43664CB00006B/200